ICT and Economic Growth

EVIDENCE FROM OECD COUNTRIES, INDUSTRIES AND FIRMS

OECD

ORGANISATION FOR ECONOMIC CO-OPERATION AND DEVELOPMENT

ORGANISATION FOR ECONOMIC CO-OPERATION AND DEVELOPMENT

Pursuant to Article 1 of the Convention signed in Paris on 14th December 1960, and which came into force on 30th September 1961, the Organisation for Economic Co-operation and Development (OECD) shall promote policies designed:

- to achieve the highest sustainable economic growth and employment and a rising standard of living in member countries, while maintaining financial stability, and thus to contribute to the development of the world economy;
- to contribute to sound economic expansion in member as well as non-member countries in the process of economic development; and
- to contribute to the expansion of world trade on a multilateral, non-discriminatory basis in accordance with international obligations.

The original member countries of the OECD are Austria, Belgium, Canada, Denmark, France, Germany, Greece, Iceland, Ireland, Italy, Luxembourg, the Netherlands, Norway, Portugal, Spain, Sweden, Switzerland, Turkey, the United Kingdom and the United States. The following countries became members subsequently through accession at the dates indicated hereafter: Japan (28th April 1964), Finland (28th January 1969), Australia (7th June 1971), New Zealand (29th May 1973), Mexico (18th May 1994), the Czech Republic (21st December 1995), Hungary (7th May 1996), Poland (22nd November 1996), Korea (12th December 1996) and the Slovak Republic (14th December 2000). The Commission of the European Communities takes part in the work of the OECD (Article 13 of the OECD Convention).

Publié en français sous le titre :

Les TIC et la croissance économique
PANORAMA DES INDUSTRIES, DES ENTREPRISES ET DES PAYS DE L'OCDE

Foreword

The 2001 OECD Ministerial report, "The New Economy: Beyond the Hype", concluded that information and communications technologies (ICT) are important, with the potential to contribute to more rapid growth and productivity gains in the years to come. Both the 2001 and 2002 OECD Ministerial meetings reiterated the importance of ICT for growth performance and requested the OECD to continue its work in this area. A request for further work on ICT and business performance was also made to the OECD in the autumn of 2001, by the US Secretary of Commerce, Mr. Evans.

This report, which responds to OECD Ministers, revisits the contribution made by ICT to economic performance using new and more recent data to assess the degree to which the findings that appeared valid at the end of 2000 remain intact. The report also examines whether the policy conclusions from the previous OECD work require adjustment in the current economic environment, and what measures OECD governments should take to seize the benefits of ICT. The findings and policy implications of the work reaffirm and elaborate those of the OECD growth study.

The report draws on work carried out in the OECD Directorate for Science, Technology and Industry, as well as work by statistical offices and research institutions in 13 OECD countries. Dirk Pilat was the principal author of the report. Important contributions were received from Alessandra Colecchia, Andrew Devlin, Frank Lee, Paul Schreyer, Colin Webb and Andrew Wyckoff. Comments were received from across the OECD Secretariat. Drafts of this report were discussed by the Committee for Industry and Business Environment and the Committee for Information, Computer and Communications Policy. Participants at these meetings provided useful comments. A grant for this work by the US Department of Commerce is gratefully acknowledged. This work could not have been completed without the contributions of statisticians and researchers in many OECD member countries. This report highlights their achievements and is dedicated to their work.

Table of Contents

List of Boxes

Main findings

The 2001 OECD Ministerial report, "The New Economy: Beyond the Hype", concluded that information and communications technology (ICT) is important and has the potential to contribute to more rapid growth and productivity gains in the years to come. Both the 2001 and 2002 OECD Ministerial meetings reiterated the importance of ICT for growth performance and requested the OECD to continue its work in this area. A specific request for further work on ICT and business performance was made to the OECD in the autumn of 2001, by the US Secretary of Commerce, Mr. Evans. This report, which responds to OECD Ministers and Secretary Evans, revisits the contribution made by ICT to economic performance using more recent data to assess the degree to which the empirical findings that appeared valid at the end of 2000 remain in tact. It draws on a range of new statistics and empirical analysis that was not available for prior OECD work. This includes new empirical analysis with official firm-level data that has been carried out through an OECD-led team of researchers and statistical offices in 13 OECD countries. The study also incorporates new evidence from official statistics on the use of ICT and e-commerce by firms, which were also not available for previous work. The report also examines whether the policy conclusions from the previous OECD work require adjustment in the current economic environment. The findings and policy implications of the work are summarised below; they reaffirm and elaborate those of the OECD Growth Study.

Empirical messages

ICT continues to have strong impacts on performance

The recent slowdown has laid to rest several myths regarding the new economy: the business cycle is not dead, stock market valuations must be realistic and backed by sound profit expectations, and the ICT sector is not immune to downturns. But this should not distract from the economic benefits that have already accompanied the spread of ICT and the continued importance of ICT for growth in the years to come. It may be too early to tell how the role of ICT in growth and productivity performance will develop in the first decade of the 21st century. Some general trends can be observed, however, that suggest that ICT will continue to be a driver of growth:

- Productivity growth in the United States, the main example of ICT-led growth and productivity improvements, has continued to be strong during the recent slowdown, suggesting that part of the acceleration in productivity growth over the second half of the 1990s was indeed structural. Productivity growth in Australia and Canada, both countries characterised by ICT-intensive growth, was also strong over the recent past.

● ICT networks have now spread throughout much of the OECD business sector, and will increasingly be made to work to enhance productivity and business performance. Technological progress in ICT goods and services is continuing at a rapid pace, driving prices down and leading to a wide range of new applications. For example, business-to-consumer e-commerce continues to gain in importance, broadband is diffusing rapidly, and activity in the telecommunications sector continues to grow. Moreover, several applications, such as broadband and e-commerce, are still in their early stages and may have a large potential for future growth.

● While ICT investment has dropped off during the recent slowdown, the release of increasingly powerful microprocessors is projected to continue for the foreseeable future, which will encourage ICT investment and support further productivity growth. Nevertheless, the level of ICT investment may well be lower than that observed prior to the slowdown, however, as the 1995-2000 period was characterised by some one-off investment peaks, *e.g.* investments related to Y2K and the diffusion of the Internet. On the other hand, some countries may still have scope for catch-up; by 2000, Japan and the European Union area invested a similar share of total investment in ICT than the United States did in 1980.

● Further technological progress in ICT production will imply a continued positive contribution of the ICT manufacturing sector to multifactor productivity (MFP) growth, notably in countries with large ICT-producing sectors such as Finland, Ireland, Japan, Korea, Sweden and the United States.

The impacts of ICT differ markedly across OECD economies

Despite the importance of ICT, there continue to be marked differences in the diffusion of ICT across OECD countries. New OECD data show that the United States, Canada, New Zealand, Australia, the Nordic countries and the Netherlands typically have the highest rates of diffusion of ICT. Many other OECD countries lag in the diffusion of ICT and have scope for greater uptake. It is likely that the largest economic benefits of ICT should be observed in countries with high levels of ICT diffusion. However, having the equipment or networks is not enough to derive economic benefits. Other factors, such as the regulatory environment, the availability of appropriate skills, the ability to change organisational set-ups, as well as the strength of accompanying innovations in ICT applications, affect the ability of firms to seize the benefits of ICT. Consequently, countries with equal ICT diffusion will not always have similar impacts of ICT on economic performance.

The measured economic impacts of ICT that have been observed thus far differ markedly across OECD countries. Three impacts of ICT on economic growth can be distinguished. First, investment in ICT adds to the capital stock that is available for workers and thus helps raise labour productivity. ICT investment accounted for between 0.3 and 0.8 percentage point of growth in GDP per capita over the 1995-2001 period. The United States, Canada, the Netherlands and Australia received the largest boost; Japan and the United Kingdom a more modest one, and Germany, France and Italy a much smaller one. Investment in software accounted for up to a third of the contribution of ICT capital.

Second, the ICT-producing sector plays an important role in some OECD countries, although it is small in most. Having an ICT-producing sector can be important, since it has been characterised by rapid technological progress and very strong demand. In Finland, Ireland and Korea, close to 1 percentage point of aggregate labour productivity growth in the 1995-2001 period was due to ICT manufacturing. In the United States, Japan and Sweden, the ICT-producing sector also contributed significantly to productivity growth.

Third, new evidence from an OECD-led consortium of researchers in 13 OECD countries demonstrates that the use of ICT throughout the value chain contributes to improved firm performance. The smart use of ICT can help firms increase their overall efficiency in combining labour and capital, or multi-factor productivity (MFP). ICT use can also contribute to network effects, such as lower transaction costs and more rapid innovation, which can improve MFP. In some countries, notably the United States and Australia, there is evidence that sectors that have invested most in ICT, such as wholesale and retail trade, have experienced more rapid MFP growth.

Firm-level studies also show that the use of ICT may help efficient firms gain market share at the cost of less productive firms, raising overall productivity. In addition, the use of ICT may help firms expand their product range, customise the services offered, or respond better to demand, i.e. to innovate. Moreover, ICT may help reduce inventories or help firms integrate activities throughout the value chain. These studies also show that ICT is part of a broader range of changes that help enhance performance. The impacts of ICT depend on complementary investments, e.g. in appropriate skills, and on organisational changes, such as new strategies, new business processes and new organisational structures. Firms adopting these practices tend to gain market share and enjoy higher productivity gains than other firms.

ICT use by firms is also closely linked to the ability of a company to adjust to changing demand and to innovate. Users of ICT often help make their investments more valuable through their own experimentation and innovation, e.g. the introduction of new processes, products and applications. Without this process of "co-invention", which often has a slower pace than technological innovation, the economic impact of ICT would be more limited. Firms that have introduced process innovations in the past are often particularly successful in using ICT; in Germany, the impact of ICT investments on output was about four times higher in firms introducing innovations as in other firms. The impacts of ICT on innovation are particularly important in services, as ICT helps in re-inventing business processes and developing new applications.

The positive impacts of ICT use on firm performance are not limited to the United States, where the macroeconomic impacts of ICT are considered largest, but are found in many OECD countries. This may be caused by several factors. First, aggregation across industries may disguise some of the impacts of ICT, as strong effects in some industries may be counterbalanced by weak effects in others. Second, the size of the measured impact at the firm level may be smaller outside the United States, as networks are often less developed and conditions for ICT to become effective may be less well established. This would lead to a smaller macroeconomic impact. Third, the impacts of ICT may be insufficiently picked up in data outside the United States, *e.g.* due to differences in the measurement of output in the services sector. Fourth, countries outside the United States may not yet have benefited from network – or spillover – effects that could create a gap between the impacts of ICT at the level of individual firms and those that are measured at the macroeconomic level. Such spillover effects could show up as benefits for downstream firms and consumers, and would thus not be picked up in the evidence for individual firms using ICT. Finally, in a large and highly competitive market such as the United States, firms investing in ICT may not always be the main beneficiaries of their investment. Consumers may extract a large part of the benefits, in the form of lower prices, better quality, improved convenience, and so on. In countries with less competition, firms might be able to extract a greater part of the returns, and spillover effects might be more limited.

ICT is no panacea

Firms may well overinvest in ICT, either in an effort to compensate for lack of skills or competitive pressure, or because they lack a clear market strategy. However, ICT is no panacea, but a technology that can be made to work to enhance business performance. Evidence for Germany shows that firms that were able to extract the benefits from ICT were those that had already successfully innovated, *i.e.* changed their products and processes. Moreover, ICT requires many other changes to make it work.

It also takes time to adapt to ICT, *e.g.* in changing organisational set-ups and worker-specific skills. Firms that adopted network technologies several years ago, notably large firms, have often already been able to make the technology work, whereas more recent adopters are still adapting their organisation, management or skills. Evidence for the United Kingdom shows that among the firms that had already adopted ICT technologies in or before 1995, over 50% purchased through electronic networks in 2000. For firms that only adopted ICT in 2000, less than 20% purchased through electronic networks in 2000. This also suggests that some of the benefits of prior ICT investments may still emerge in the future.

The impacts of ICT are affected by differences in the business environment

The firm-level evidence also confirms that there are important cross-country differences in firms' use of ICT. New firms in the United States seem to experiment more with business models than those in other OECD countries; they start at a smaller scale than European firms, but grow much more quickly when successful. This may be linked to less aversion to risk in the United States, linked to its financial system, which provides greater opportunities for risky financing to innovative entrepreneurs. Moreover, low regulatory burdens enable US firms to start at a small scale, experiment, test the market and their business model, and, if successful, expand rapidly. Moreover, if they do not succeed, the costs of failure are relatively limited. In contrast, firms in many other OECD countries are faced with high entry and exit costs. In a period of rapid technological change, greater scope for experimentation may enable new ideas and innovation to emerge more rapidly, leading to faster technology diffusion.

The empirical analysis also points to several other factors that affect firms' investment and the diffusion of ICT. These can be divided into several categories:

- Factors related to the *direct costs of ICT, e.g.* the costs of ICT equipment, telecommunications or the installation of an e-commerce system.

- *Costs and implementation barriers related to enabling factors, e.g.* linked to the availability of know-how or qualified personnel, the scope for organisational change, or the scale of accompanying innovations in ICT applications.

- *Factors related to risk and uncertainty, e.g.* the security of doing business online or the uncertainty of payments, delivery and guarantees online.

- *Factors related to the nature of the businesses.* ICT is a general-purpose technology, but is more appropriate for some activities than for others. ICT may not fit in all contexts and specific applications, such as electronic commerce, may not be suited to all business models.

- *Factors related to competition.* A competitive environment is more likely to lead a firm to invest in ICT, as a way to strengthen performance and survive, than a more sheltered environment. Competition also puts downward pressure on the costs of ICT, thus promoting its diffusion.

Policy implications

The analysis confirms and reaffirms the findings of the OECD work on growth, and its policy conclusions as regards ICT diffusion and economic growth. The main policy conclusions that can be drawn are:

1. *Competition in ICT goods and services needs to be strengthened.* Competition in ICT goods and services requires attention, as continued technological change is creating new challenges to competition in many markets.

2. *The business environment needs to be improved.* This includes, *inter alia*, having an environment that provides access to finance, allows firms to change the organisation of functions and tasks, helps workers acquire the skills they need in a rapidly changing global environment, and promotes good management practices. Rigid regulations of product and labour markets that impede reorganisation or competition between firms also need to be addressed. The experience of countries such as Australia shows that structural reform is key in harnessing the new dynamism that is associated with ICT. Firm creation also needs to be fostered. Experimentation and competition are key in selecting those firms that seize the benefits of ICT and in making them flourish and grow. In the current time of rapid technological change, greater scope for experimentation may enable new ideas and innovation to emerge more rapidly, leading to faster technology diffusion. Barriers to the entry, exit and growth of firms therefore need to be addressed. Moreover, competition needs to be strengthened. Competition not only helps lower the costs of ICT products and services, which fosters diffusion, it also strengthens pressures on firms to improve performance and change conservative attitudes.

3. *Security and trust need to be boosted.* Concerns on security, privacy and authentification continue to affect the uptake and use of ICT and should remain a priority for policy.

4. *Barriers to the effective use of ICT in services require attention.* Sector-specific regulations reduce the development of new ICT applications and limit the capability of firms to seize the benefits of ICT. Further reform of regulatory structures is needed to promote competition and innovation, and to reduce barriers and administrative rules for new entrants and start-ups.

5. *Innovation needs to be harnessed to draw the benefits from ICT.* ICT is closely linked to the ability of firms to innovate, *i.e.* introduce new products, services, business processes, and applications. Firms that have already innovated achieve much better results from ICT than those that have never innovated. Policies to harness the potential of innovation, as outlined in the OECD Growth Study, are thus of great importance in seizing the benefits of ICT. To strengthen innovation, policy needs to give greater priority to fundamental research, improve the effectiveness of public R&D funding and promote the flow of knowledge between science and industry.

Introduction

The 2001 OECD Ministerial report "The New Economy: Beyond the Hype" concluded that information and communications technology (ICT) is important and has the potential to contribute to more rapid growth in the future. Both the 2001 and 2002 OECD Ministerial meetings reiterated the importance of ICT for growth performance and requested the OECD to continue its work on ICT. A specific request for further work on ICT and business performance was also made to the OECD in the autumn of 2001, by the US Secretary of Commerce, Mr. Evans. This report responds to these requests and examines whether ICT is still important now the hype of the new economy is over. It provides new evidence on the factors that affect the diffusion of ICT across OECD countries, the economic impacts of ICT, the factors that influence these impacts, and the policies that can help countries in seizing the benefits from ICT.

This study differs from previous OECD work on growth and the role of ICT as it considers a range of questions that were not explicitly addressed in the previous work by the OECD. For example, why have some OECD countries invested more in ICT than others? What characterises firms that adopt ICT? Which technologies are they using and for which purpose? What factors help firms in seizing the benefits from ICT? How precisely does ICT affect firm performance, *e.g.* in strengthening productivity and in increasing market shares? And what policies are key in helping firms seize the benefits of ICT?

Many of these questions can not easily be examined with the macroeconomic and sectoral data that were used in previous OECD work. Firm-level data are often necessary, since they allow interactions at the firm level to be examined. For example, the role of ICT in helping firms gain market share can only be analysed with firm-level data, as can the role of organisational change. Studies drawing on firm-level evidence can thus contribute to a better understanding of the drivers of economic performance, *e.g.* the interaction between ICT, human capital, organisational change and innovation, and thus to better, evidence-based, policy making.

The report also draws on a range of new data. First, it draws on new empirical analysis with official firm-level data that has been carried out through an OECD-led team of researchers and statistical offices in 13 OECD countries, thus complementing the sectoral and aggregate analysis.* Second, the study incorporates new evidence from official statistics on the use of ICT and e-commerce by firms, which were not available for previous work. Third, it draws, to the extent possible, on the latest available data to examine the contribution of ICT to growth performance in recent years.

Chapter 1 of this report examines the diffusion of ICT across OECD countries. It uses official statistics, which may differ substantially from previously published private estimates of ICT diffusion. It shows that ICT networks continue to diffuse throughout the OECD area, even during a period of slower growth. However, large differences in the uptake of technologies persist across the OECD, both between and within OECD countries. Cost differentials and structural differences are among the factors explaining these differences, as is the state of the business environment in different OECD countries.

Chapter 2 provides evidence on the impact of ICT at the macroeconomic and sectoral level, updating and extending previous OECD work. It shows that ICT investment has contributed to growth and labour productivity in all OECD countries for which data are available, but more in the United States than in any other OECD country. Moreover, rapid technological progress in the ICT-producing industry has contributed to rapid productivity growth in some OECD countries, notably the United States, Finland, Ireland, Korea and Sweden. In a few countries, notably the United States and Australia, there is also evidence that sectors that have invested much in ICT, for example wholesale and retail trade, have experienced more rapid productivity growth. These impacts of ICT on productivity have not disappeared with the recent slowdown; part of the acceleration in aggregate productivity growth in certain OECD countries, notably the United States, is structural and could continue in the years to come.

Chapter 3 provides evidence on the contribution of ICT use to business performance, based on detailed firm-level studies. It demonstrates that the use of ICT indeed contributes to improved business performance, but only when it is complemented by other investments and actions at the firm level, such as changes in the organisation of work and changes in workers skills. The chapter also shows that ICT is no panacea; investment in ICT does not compensate for poor management, lack of skills, lack of competition, or a low ability to innovate. Not all firms will therefore succeed in generating returns from their ICT investments; many will fail. In addition, drawing the benefits from ICT investment takes time. Moreover, there are important cross-country

---

* These countries are: Australia, Canada, Denmark, Finland, France, Germany, Italy, Japan, Netherlands, Sweden, Switzerland, United Kingdom and United States. See Box 3.2 for details.

differences in firms' use of ICT. Firms in the United States are characterised by a much higher degree of experimentation in their use of ICT than European firms; they take higher risks and opt for potentially higher outcomes.

Chapter 4 draws implications from the empirical evidence presented in Chapters 1 to 3 for policy makers. It argues that governments should reduce unnecessary costs and regulatory burdens on firms to create a business environment that promotes productive investment. This involves policies to enable organisational change, to strengthen education and training systems, to encourage good management practices, and to foster innovation, *e.g.* in new applications, that can accompany the uptake of ICT. Moreover, policy should foster market conditions that reward the successful adoption of ICT; a competitive environment is key for this to happen. Governments will also need to work with business and consumers to shape a regulatory framework that strengthens confidence and trust in the use of ICT, notably electronic commerce. Policies to foster growth in services are important too, as ICT offers a new potential for growth in the service sector, providing that regulations that stifle change are adjusted or removed. Finally, the report reaffirms the importance of economic and social fundamentals as the key to lasting improvements in economic performance. A short set of conclusions completes the report.

ISBN 92-64-10128-4
ICT and Economic Growth: Evidence from OECD Countries,
Industries and Firms
© OECD 2003

Chapter 1

The diffusion of ICT in OECD economies

Abstract.   *This chapter examines the diffusion of ICT across OECD countries. It uses official statistics, which may differ substantially from previously published private estimates of ICT diffusion. It shows that ICT networks continue to diffuse throughout the OECD area, even during a period of slower growth. However, large differences in the uptake of technologies persist across the OECD, both between and within OECD countries. Cost differentials and structural differences are among the factors explaining these differences, as is the state of the business environment in different OECD countries.*

The state of ICT diffusion

The economic impact of ICT is closely linked to the extent to which different ICT technologies have diffused across OECD economies. This is partly because ICT is a network technology; the more people and firms that use the network, the more benefits it generates. The diffusion of ICT currently differs considerably between OECD countries, since some countries have invested more or have started earlier to invest in ICT than other countries.

A core indicator of ICT diffusion is the share of ICT in investment. Investment in ICT establishes the infrastructure for the use of ICT (the ICT networks) and provides productive equipment and software to businesses. While ICT investment has accelerated in most OECD countries over the past decade, the pace of that investment differs widely. The data show that ICT investment rose from less than 15% of total non-residential investment in the early 1980s, to between 15% and 30% in 2001. In 2001, the share of ICT investment was particularly high in the United States, the United Kingdom, Sweden, the Netherlands, Canada and Australia (Figure 1.1). ICT investment in many European countries was substantially lower than in the United States.

The high growth of ICT investment has been fuelled by a rapid decline in the relative prices of computer equipment and the growing scope for the application of ICT. Due to rapid technological progress in the production of key ICT technologies, such as semi-conductors, and strong competitive pressure in their production,[1] the prices of key technologies have fallen by between 15 and 30% annually, making investment in ICT attractive to firms. The benefits of lower ICT prices have been felt across the OECD, as both firms investing in these technologies and consumers buying ICT goods and services have benefited from lower prices. The lower costs of ICT are only part of the picture; ICT is also a technology that offers large potential benefits to firm, e.g. in enhancing information flows and productivity. Chapter 2 examines the impact of ICT investment on economic growth in more detail.

A second determinant of the economic impacts associated with ICT is the size of the ICT sector, i.e. the sector that produces ICT goods and services. Having an ICT-producing sector can be important, since ICT-production has been characterised by rapid technological progress and has been faced with very strong demand. The sector has therefore grown very fast, and made a large contribution to economic growth, employment and exports. Moreover, having a strong ICT sector may help firms that wish to use ICT, since the close

Figure 1.1. **ICT investment in selected OECD countries**

As a percentage of non-residential gross fixed capital formation, total economy

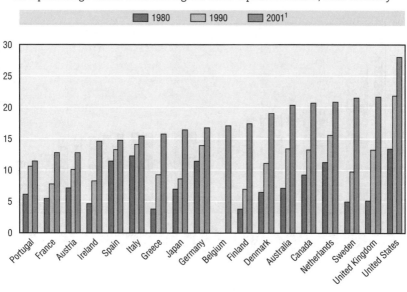

Note: Estimates of ICT investment are not yet fully standardised across countries, mainly due to differences in the capitalisation of software in different countries. See Ahmad (2003).
1. Or latest available year.
Source: OECD, Database on Capital Services.

proximity of producing firms might have advantages when developing ICT applications for specific purposes. In addition, having a strong ICT sector should also help generate the skills and competencies needed to benefit from ICT use. And it could also lead to spin-offs, as in the case of Silicon Valley or in other high technology clusters. Having an ICT sector can thus support growth, although previous OECD work has shown that it is not a prerequisite (OECD, 2001a).

In most OECD countries, the ICT sector is relatively small, although it has grown rapidly over the 1990s.[2] In 2000, value added in the ICT sector represented between 4% and 17% of business sector value added (Figure 1.2). Moreover, about 6-7% of total business employment in the OECD area can be attributed to ICT production. Trade in ICT has also grown very rapidly, growing from just over 12% of total trade in 1990, to almost 18% in 2000 (OECD, 2002a). Chapter 2 examines the contribution of the ICT-producing sector to economic performance in more detail.

A third factor that affects the impact of ICT in different OECD countries is the distribution of ICT across the economy. In contrast to Solow's famous remark, "you see computers everywhere but in the productivity statistics"

Figure 1.2. **Share of the ICT sector in value added, non-agricultural business sector, 2000**

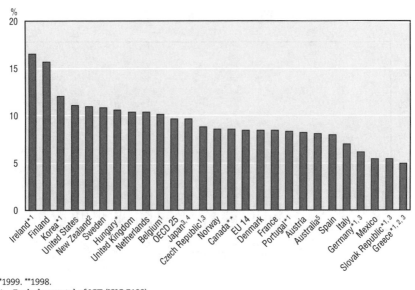

*1999. **1998.
1. Excludes rental of ICT (ISIC 7123).
2. Includes postal services.
3. Excludes ICT wholesale (ISIC 5150).
4. Includes only part of computer-related activities.
5. 2000-2001.

Source: OECD (2002a), *Measuring the Information Economy, www.oecd.org/sti/measuring-infoeconomy*

(Solow, 1987), computers are, in fact, heavily concentrated in the service sector. Figure 1.3 shows evidence for the United States. It shows the share of the total stock of equipment and software that is accounted for by IT equipment and software (excluding communications equipment). The graph shows that more than 30% of the total stock of equipment and software in legal services, business services and wholesale trade consists of IT and software. Education, financial services, health, retail trade and a number of manufacturing industries (instruments and printing and publishing) also have a relatively large share of IT capital in their total stock of equipment and software. The average for all private industries is just over 11%. The goods-producing sectors (agriculture, mining, manufacturing and construction) are much less IT-intensive; in several of these industries less than 5% of total equipment and software consists of IT.

The relative distribution of ICT investment across sectors for other OECD countries is not very different for other OECD countries (Van Ark *et al.*, 2002; Pilat *et al.*, 2002); services sectors such as wholesale trade and financial services are typically the most intensive users of ICT.[3] This may suggest that any impacts on economic performance might be more visible in the services

Figure 1.3. **Information technology as a percentage of all stock of equipment and software, United States, 2001**

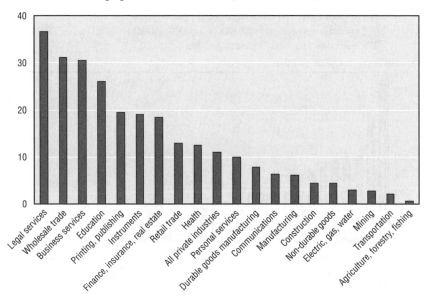

Source: Bureau of Economic Analysis, US Department of Commerce, Fixed Assets Tables, *www.bea.doc.gov/*

sectors than in other parts of the economy. Nevertheless, ICT is commonly considered to be a general-purpose technology, as all sectors of the economy use information in their production process, which implies that all sectors might be able to benefit from the use of ICT. Chapters 2 and 3 return to the sectoral aspects of ICT use.

The distribution of ICT also differs according to the size of firms. Smaller firms are typically less ICT-intensive than large firms. This is partly because large firms have more scope for improving communication flows within the firm, *e.g.* establishment of intra-firm networks, or by outsourcing different tasks, *e.g.* creation of extranets. But large firms also invest more in ICT than small firms since ICT investment is risky and uncertain, which may be more difficult to bear for small firms. This may obviously imply that the impacts of ICT use could be larger for large firms than for small firms. Chapter 3 provides further evidence on this issue.

One further indicator that points to the uptake of ICT is the number of secure servers in each country (Figure 1.4). This measures the number of servers that use secure software for purchasing goods and services or transmitting privileged information over the Internet, *e.g.* credit card details. It therefore provides an indication of the development of Internet-based activity

Figure 1.4. **Internet commerce measured by the number of secure Web servers**

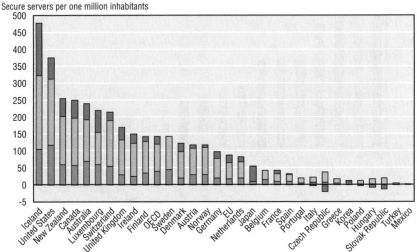

Source: OECD and Netcraft (*www.netcraft.com*), December 2002.

in different countries, notably the increased security of such activity. By July 2002, over 65% of all secure servers were located in the United States. The United Kingdom had the second largest number of secure servers, accounting for about 6% of the OECD total. Large OECD countries such as Japan, France and Italy had relatively few secure servers, especially when measured on a per capita basis. Among smaller countries, Iceland, New Zealand and Australia have a relatively large number of secure servers. The diffusion of secure servers continued unabated between July 2001 and July 2002, even though ICT investment slowed down during the period.

Another, more accurate, indicator of the development of electronic commerce is the proportion of businesses that use the Internet to purchases and sales (Figure 1.5). This is available for fewer countries, but roughly confirms the findings of the previous graph, with a large number of firms using the Internet for sales or purchases in the Nordic countries (Denmark, Finland, Norway and Sweden) as well as in Australia, the Netherlands and New Zealand. In contrast, only few firms in Greece, Italy, Portugal and Spain use the Internet for sales or purchases, even if many are connected to the Internet.

Figure 1.5. **Proportion of businesses using the Internet for purchases and sales, 2001**[1]

Percentages of businesses with ten or more employees

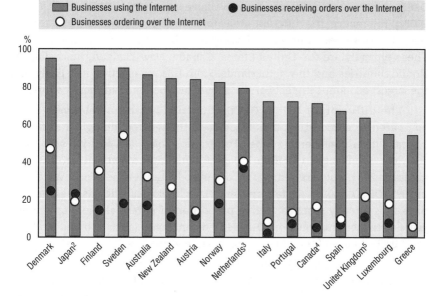

Note: The Eurostat survey is based on a selection of industries which changes slightly across countries. See source for detail.

1. Beginning of 2001 for Internet use, purchases and sales refer to 2000 except for Canada where they refer to 2001, and Denmark and Norway where Internet use refers to 2002 and purchases and sales to 2001.
2. All businesses with 50 and more employees.
3. Use, orders received and placed refer to Internet and other computer mediated networks.
4. All businesses.
5. Orders received and made over the Internet and other computer mediated networks.

Source: OECD, *Measuring the Information Economy*, 2002, based on Eurostat, E-Commerce Pilot Survey 2001.

Monetary estimates of the importance of electronic commerce are also available for several OECD countries, although these are not yet entirely comparable, depending on the definition used and the coverage of different sectors. The available data suggest that electronic commerce is growing, albeit more slowly than originally envisaged, but still accounts for a relatively small proportion of overall sales. For the few countries that currently measure the value of Internet or electronic sales, total Internet sales in 2000-01 ranged between 0.2% and 2% of total sales. In the fourth quarter of 2002, 1.65% of all retail sales in the United States were carried out through computer-mediated networks, up from 1.3% in the fourth quarter of 2001. Sales via EDI (electronic data interchange) are generally higher than sales via the Internet, with almost all countries reporting EDI sales to be at least twice as high as Internet sales.

In 2000, electronic sales (including those over all computer-mediated networks) were over 13% in Sweden.

There are many other indicators that point to the role of ICT in different OECD economies, most of which are available in a separate OECD study (OECD, 2002a). In practice, the different indicators are closely correlated and tend to point to the same countries as having the highest rate of diffusion of ICT. These typically are the United States, Canada, New Zealand, Australia, the Nordic countries and the Netherlands. From this perspective, it is likely that the largest economic impacts of ICT should also be found in these countries.

The diffusion of ICT in OECD countries has been relatively rapid compared to some other technologies, although technological diffusion typically takes considerable time.[4] For example, over 90% of firms with more than ten employees in Denmark, Japan, Finland and Sweden had Internet access in 2001, only six years after the introduction of the World Wide Web in 1995 (OECD, 2002a). Certain recent ICT technologies (such as the Internet) have thus already reached a large proportion of potential users only a few years after their introduction. Other ICT technologies (such as broadband) are in an earlier stage of the diffusion process, however.

The diffusion of ICT continues across OECD economies, despite the current slowdown. The share of ICT investment in total capital formation grew rapidly until 2000, and remained at a high share of investment even in 2001 and 2002, suggesting that ICT investment has not been affected disproportionally by the slowdown compared with other types of investment. Evidence for the United States shows that ICT investment was among the first areas of investment to recover in 2002 (BEA, 2003). The continued diffusion of ICT can also be observed in other areas (Figure 1.6). For example, the number of secure servers continued to grow between July 2001 and December 2002, as did the number of broadband subscribers. This rose from 33 million by the end of 2001, to 42 million by June 2002, and to more than 55 million by the end of 2002. This is not to suggest that the ICT sector is not going through a slowdown. However, large ICT networks are now in place throughout the business sector. These will have to be maintained and updated, and will increasingly be made to work and generate economic returns.

Factors affecting the diffusion of ICT

Why is the diffusion of ICT so different across OECD countries? Previous OECD work already noted several factors, such as lack of relevant skills, lack of competition, or high costs (OECD, 2001a). From a firm's perspective, high costs are particularly important, as they affect the possible returns that a firm can extract from their investment. Firms do not only incur costs in acquiring new technologies, but also in making it effective in the workplace, and in using the

Figure 1.6. **Diffusion of key ICT technologies**

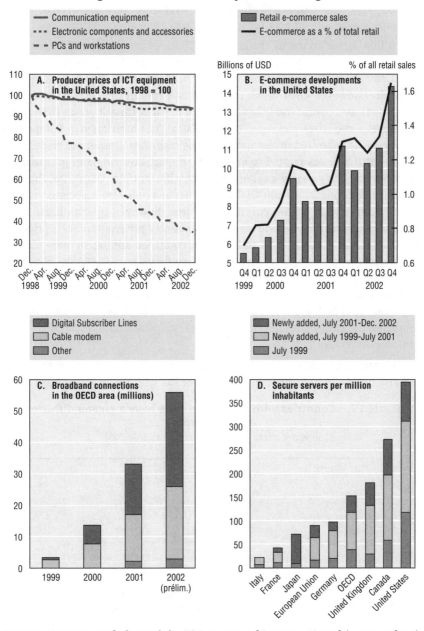

Source: OECD, US Bureau of Labor Statistics, US Department of Commerce, Netcraft (www.netcraft.com).

technologies on a daily basis. Costs related to personnel, telecommunication charges and organisational change are therefore also important. Some evidence is available on how these factors may have affected diffusion.

A first factor concerns the costs of ICT hardware. Since ICT investment goods are traded internationally, their prices should not vary too much across OECD countries. Evidence from international price comparisons suggests otherwise, however. Over much of the 1990s, firms in the United States and Canada enjoyed considerably lower costs of ICT investment goods than firms in European countries and Japan (OECD, 2001a). The high costs in Europe and Japan may have limited investment in these countries. Barriers to trade, such as non-tariff barriers related to standards, import licensing and government procurement, may partly explain the cost differentials (OECD, 2002b). The higher price levels in certain OECD countries may also be associated with a lack of competition within countries. In time, however, international trade and competition should erode these cross-country price differences; prices of ICT investment goods in 1996 in European countries and Japan were already much closer to those in the United States than they were in 1993. By 1999, they had come down further across the OECD (OECD, 2002c). The costs of telecommunication are also important, as they affect the use of ICT in networks. This is illustrated by the take-up of the Internet in different OECD countries. Countries with lower access costs typically have a higher take-up of the Internet (Figure 1.7).

Evidence on perceived barriers to the uptake of selected ICT technologies can also be drawn from firm-level surveys of ICT use. These surveys ask firms

Figure 1.7. **Countries with low access costs have a greater diffusion of the Internet**

Internet hosts per 1 000 inhabitants, July 2001

Average price for 20 hours Internet access 1995-2000, in PPP dollars

Source: OECD (*www.oecd.org/dsti/sti/it/cm*) and Telcordia Technologies (*www.netsizer.com*).

Figure 1.8. **Perceived barriers to Internet access and use
in the business sector, 2000**

Percentage of businesses using a computer with 10 or more employees

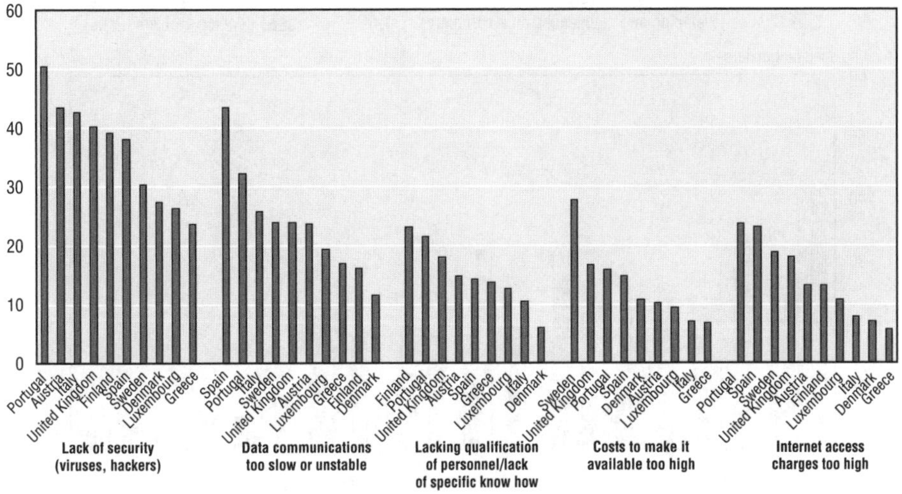

and consumers about the barriers they face in using the Internet and electronic commerce. While these surveys currently only cover a limited number of OECD countries, some interesting patterns emerge (OECD, 2002a). As regards Internet access, lack of security and slow or unstable communications were considered the key problems in European countries (Figure 1.8). Other problems, such as lack of know-how or personnel, high costs of equipment or Internet access, were considered less of a problem. These barriers also differ by the size of firms; large firms tend to face fewer problems in getting qualified personnel or know-how than small firms. However, large firms tend to regard security issues as a more important barrier than small firms, perhaps because large firms use the Internet more actively than small firms. These barriers may also differ by activity; the perceived benefits of Internet use vary considerably across activities (and also differ across countries). Moreover, there is a considerable difference in the perceived barriers of firms that already use the Internet, and those that are considering its use. Evidence for Norway and the United Kingdom shows that "lack of security" is a much more important barrier for firms already using the Internet than for non-users, showing that the perception of barriers may change once a technology is actually used.

Survey information on the barriers to Internet commerce provides further insights (Figure 1.9). These suggest that legal uncertainties (uncertainty over payments, contracts, terms of delivery and guarantees)

Figure 1.9. **Barriers to Internet commerce faced by businesses, 2000**

Percentage of businesses using a computer with 10 or more employees

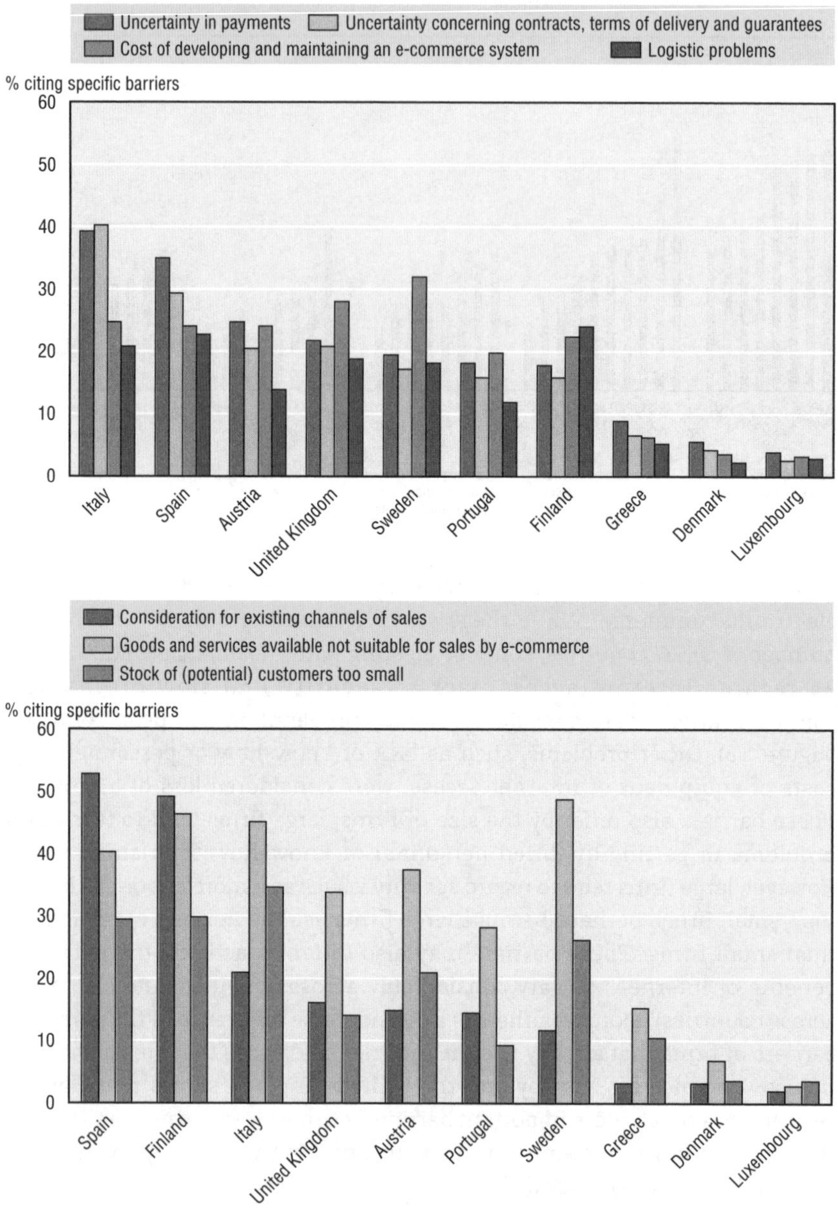

Source: OECD (2002a), *Measuring the Information Economy*, based on Eurostat, E-commerce Pilot Survey.

remain important in several countries. Business-to-consumer transactions are typically hampered by concerns about security of payment, the possibility of redress in the online environment and privacy of personal data. For business-to-business transactions, the security and reliability of systems that can link all customers and suppliers are often considered more important. Issues of system security and reliability are a major concern in Japan; almost one out of every two Japanese businesses rated viruses as the major reason for not using the Internet (Tachibana, 2000). Cost considerations remain an important issue for businesses in several countries, while logistic problems were also cited frequently.

Commercial factors were also cited by many businesses as a factor in not taking up Internet commerce. Many businesses in Finland and Spain found that Internet commerce would threaten existing sales channels. Existing transaction models or strong links with customers and suppliers along the value chain may discourage businesses from introducing new sales models. In many cases, the goods and services on offer by a particular firm were not considered suitable for Internet commerce. In Canada, among businesses that did not buy or sell over the Internet, 56% believed that their goods or services did not lend themselves to Internet transactions; 36% preferred to maintain their current business model. And firms in several countries, notably Italy, considered the market too small. Some of these considerations differ by the size of firm and the activity; large firms found logistical barriers more important than small firms did. However, barriers related to Internet payments and the costs of setting up Internet commerce did not differ in a consistent manner across OECD countries. There also differences across activities; many firms in real estate and hotels and restaurants did not consider their products and services suitable for Internet commerce, whereas only few firms in the financial sector considered this to be the case.

More elaborate analysis of this survey evidence provides further insights in the factors explaining ICT uptake. Using recent data for Switzerland, Hollenstein (2002) finds that a range of factors play a role in ICT uptake. This includes anticipated benefits of ICT adoption (*e.g.* improved customer orientation and lower costs) as well as the costs and obstacles to adoption (*e.g.* investment costs and knowledge gaps). But other factors also play a role, such as the firm's ability to absorb knowledge (linked to human capital and experience with innovation), whether the firm has experience with related technologies, as well as international competitive pressures. Sectoral differences also played an important role. Moreover, the introduction of new work practices was found to favour the adoption of ICT.

The survey evidence outlined above already suggests that the broader business environment may play a role in firm's decision to adopt ICT. This is further illustrated in Figures 1.10 and 1.11. While not demonstrating causality,

Figure 1.10. **Countries that had strict product market regulations in 1998 had lower ICT investment**

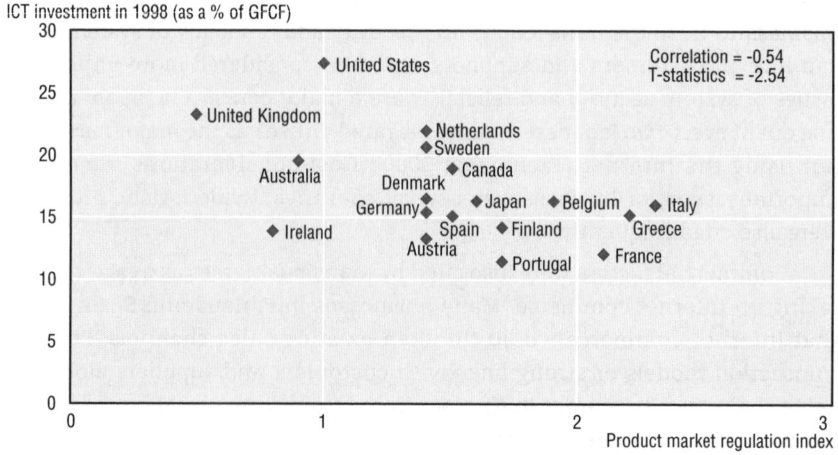

ICT investment in 1998 (as a % of GFCF)

Correlation = -0.54
T-statistics = -2.54

Product market regulation index

Notes: The scale of indicators is 0-6, from least to most restrictive. Based on the situation in or around 1998. The components are weighted to show their relative importance in the overall indicator. Since 1998, many countries have implemented reforms in product markets.

Source: ICT investment from sources quoted in Figure 1.1; regulations from Nicoletti *et al.* 1999.

Figure 1.11. **Countries with strict employment protection legislation in 1998 had lower ICT investment**

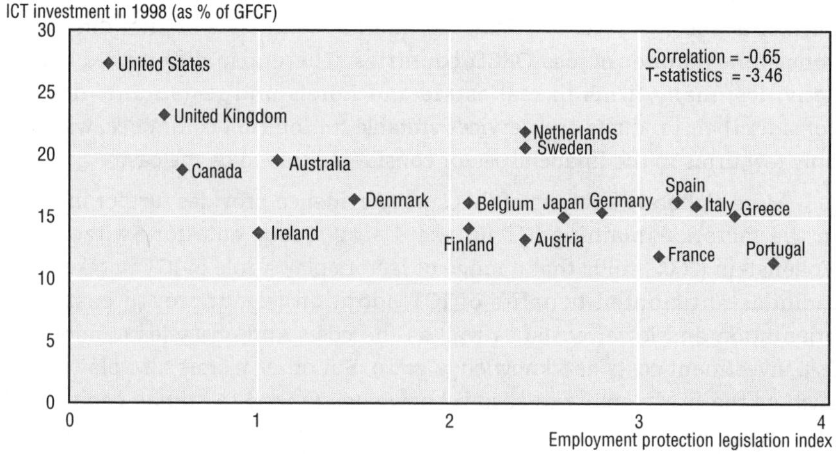

ICT investment in 1998 (as % of GFCF)

Correlation = -0.65
T-statistics = -3.46

Employment protection legislation index

Notes: The scale of indicators is 0-6, from least to most restrictive. Based on the situation in or around 1998. The components are weighted to show their relative importance in the overall indicator. Since 1998, many countries have implemented reforms in employment protection legislations.

Source: ICT investment from sources quoted in Figure 1.1; regulations from Nicoletti *et al.* 1999.

Figure 1.10 shows that there is a link between ICT investment as a share of total capital formation in 1998 and product market regulations, as measured by an OECD index of the state of these regulations in 1998. The graph shows that countries that had a high level of regulation in 1998 had lower shares of investment in ICT than countries with low degrees of product market regulation. This may be because product market regulations can limit competition. Competition is important in spurring ICT investment as it forces firms to seek ways to strengthen performance relative to competitors. Moreover, competition may help lower the costs of ICT, which stimulates diffusion. Product market regulations may also limit firms in the ways that they can extract benefits from their use of ICT, as it may reduce the incentives for firms to innovate and develop new ICT applications (OECD, 2002d). For example, product market regulations may limit firms' ability to extend beyond traditional sectoral boundaries.

Figure 1.11 shows the link between ICT investment and an index of employment protection legislation for 1998. The correlation between levels of ICT investment and labour market regulations may be related to the organisational factors that are required to make ICT work; if firms cannot adjust their workforce or organisation and make ICT effective within the firm, they may decide to limit investment or relocate activities. These links between regulations and ICT investment have been confirmed through econometric analysis; Gust and Marquez (2002) find that regulations impeding workforce reorganisations and competition between firms hinder investment in ICT. Bartelsman et al. (2002), Bartelsman and Hinloopen (2002) and Devlin (2003) also confirm these findings. Chapter 3 will return to this issue.

Diffusion in the OECD area – some conclusions

This chapter has shown that ICT has diffused rapidly across OECD countries, and is continuing to spread despite the recent slowdown. However, large cross-country differences persist, also across firms and activities within countries. The United States, Canada, New Zealand, Australia, the Nordic countries and the Netherlands typically have the highest rate of diffusion of ICT. From this perspective, it is likely that the largest economic impacts of ICT should also be found in these countries. However, previous studies have shown that having the equipment or networks is not enough to derive economic impacts. Other factors play a role and countries with equal rates of diffusion of ICT will not necessarily have similar impacts of ICT on economic performance. In addition, it is possible to overinvest in ICT and some studies suggest that firms have sometimes over-invested in ICT in an effort to compensate for poor performance.

The chapter has pointed to several factors affecting the diffusion of ICT, namely:

- Factors related to the direct costs of ICT, *e.g.* the costs of ICT equipment, telecommunications or the installation of an e-commerce system.

- Costs and implementation barriers related to enabling factors, *e.g.* linked to the availability of know-how or qualified personnel, or the scope for organisational change.

- Factors related to risk and uncertainty, *e.g.* the security of doing business online or the uncertainty of payments, delivery and guarantees online.

- Factors related to the nature of the businesses. ICT is a general purpose technology, but is more appropriate for some activities than for others. ICT may not fit in all contexts and specific technologies, such as electronic commerce, may not be suited to all business models.

- Factors related to competition. A competitive environment is more likely to lead a firm to invest in ICT, as a way to strengthen performance and survive, than a more sheltered environment. Moreover, competition puts downward pressure on the costs of ICT.

These categories point to some areas that are relevant for policy development, most of which have already been the subject of OECD work over the past years. For example, measures to increase competition can help bring down costs, effective labour market and education policies may help reduce skill shortages, and risk and uncertainty may be tackled by the development of a well designed regulatory framework. Later chapters will return to these areas in more detail.

Notes

1. Aizcorbe (2002) shows that part of the decline in the prices of Intel chips can be attributed to a decline in Intel's mark-ups over the 1990s, which points to stronger competition.

2. These estimates are based on the OECD definition of the ICT sector. See OECD (2002a).

3. Health and education are also intensive ICT users but are ignored here as their output is difficult to measure.

4. Technological diffusion often follows an S-shaped curve, with slow diffusion when a technology is new and expensive, rapid diffusion once the technology is well established and prices fall, and slow diffusion once the market is saturated.

ISBN 92-64-10128-4
ICT and Economic Growth: Evidence from OECD Countries,
Industries and Firms
© OECD 2003

Chapter 2

The contribution of ICT to growth

Abstract. *This chapter provides evidence on the impact of ICT at the macroeconomic and sectoral level. It shows that ICT investment has contributed to growth and labour productivity in all OECD countries for which data are available, but more in the United States than in any other OECD country. Moreover, rapid technological progress in the ICT-producing industry has contributed to rapid productivity growth in some OECD countries, notably the United States, Finland, Ireland, Korea and Sweden. In a few countries, notably the United States and Australia, there is also evidence that sectors that have invested much in ICT, for example wholesale and retail trade, have experienced more rapid productivity growth. These impacts of ICT on productivity have not disappeared with the recent slowdown; part of the acceleration in aggregate productivity growth in certain OECD countries, notably the United States, is structural and could continue in the years to come.*

W hat precisely are the impacts that ICT can have on business performance and growth? In most analysis of economic growth, three effects are distinguished. First, as a capital good, investment in ICT contributes to overall capital deepening and therefore helps raise labour productivity. Second, rapid technological progress in the production of ICT goods and services may contribute to more rapid multifactor productivity (MFP) growth in the ICT-producing sector. And third, greater use of ICT may help firms increase their overall efficiency, and thus raise MFP. Moreover, greater use of ICT may contribute to network effects, such as lower transaction costs and more rapid innovation, which will improve the overall efficiency of the economy, i.e. MFP. This chapter discusses evidence from aggregate and sectoral data. The next chapter will examine evidence from firm-level studies.

The impact of investment in ICT

Evidence on the role of ICT investment is primarily available at the macroeconomic level, e.g. from Colecchia and Schreyer (2001) and Van Ark et al. (2002a). Both studies show that ICT has been a very dynamic area of investment, due to the steep decline in ICT prices which has encouraged investment in ICT, at times shifting investment away from other assets. While ICT investment accelerated in most OECD countries, the pace of that investment differed widely (Figure 2.1).

For the countries for which data are available, growth accounting estimates show that ICT investment typically accounted for between 0.3 and 0.8 percentage points of growth in GDP per capita over the 1995-2001 period (Figure 2.2). The United States, Australia, the Netherlands and Canada received the largest boost; Japan and United Kingdom a more modest one, and Germany, France and Italy a much smaller one. Software accounted for up to a third of the overall contribution of ICT investment to GDP growth in OECD countries (see also Colecchia and Schreyer, 2001, and Van Ark et al., 2002a).[1]

The results of these cross-country studies have been confirmed by many studies for individual countries, which are summarised in Table 2.1. National studies may differ from the results shown in Figure 2.2, due to differences in measurement. France and the United States, for instance, use specially designed "hedonic" deflators for computer equipment: these deflators adjust prices for key quality changes induced by technological progress, like higher processing speed and greater disk capacity. They tend to show faster declines

Figure 2.1. **The share of investment in ICT in total GDP**
Percentages

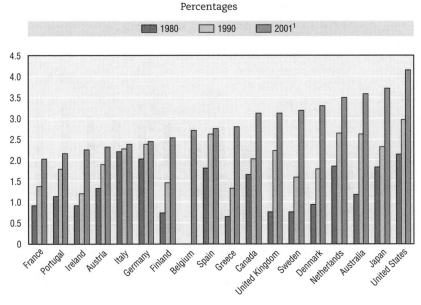

1. Or latest available year.
Source: OECD estimates based on Database on Capital Services.

in computer prices than conventional price indexes, and that means more rapid growth. As a result, countries that use hedonic indexes are likely to record faster real growth in investment and production of information and communications technology (ICT) than countries that do not use them. This faster real growth will translate into a larger contribution of ICT capital to growth performance.[2] The method used in Figure 2.2 and in the work by Colecchia and Schreyer (2001) and Van Ark *et al.* (2002) adjusts for these differences. They are therefore more comparable than the results of individual national studies. Nevertheless, the national studies typically show the same countries as experiencing a large impact of ICT investment on growth, *e.g.* Australia, Canada, Korea, the United Kingdom and the United States.

The impact of ICT investment on economic growth has not ended with the recent slowdown. While ICT investment has slowed down over the past year, technological progress in the production of computers, *i.e.* the release of increasingly powerful computer chips, is projected to continue for the foreseeable future.[3] Technological progress is also continuing at a rapid pace in other ICT technologies, such as communications technologies. This implies that quality-adjusted ICT prices will continue to decline, thus encouraging ICT investment and further productivity growth.[4] The level of ICT investment is likely to be lower than that observed prior to the slowdown, however, in

Figure 2.2. **The contribution of investment in ICT capital to GDP growth**

Percentage points contribution to annual average GDP growth, total economy

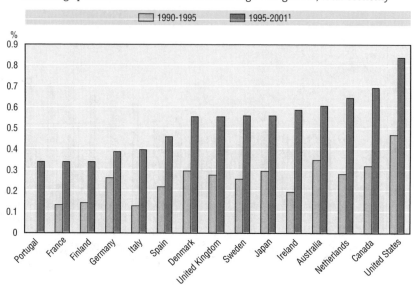

1. Or latest available year, i.e. 1995-2000 for Denmark, Finland, Ireland, Japan, Netherlands, Portugal and Sweden.

Source: OECD estimates based on Database on Capital Services. See Schreyer *et al.* (2003) for methodological details.

particular in the United States, as the 1995-2000 period was characterised by some one-off investment peaks, *e.g.* investments related to Y2K and the diffusion of the Internet (McKinsey, 2001; Gordon, 2003).

The role of ICT-producing and ICT-using sectors

Evidence on the impact of ICT can also be found from sectoral data, notably in the relative contributions of ICT-producing and ICT-using sectors to overall growth performance. The ICT-producing sector is of particular interest for several countries, as it has been characterised by very high rates of productivity growth, providing a considerable contribution to aggregate performance. Figure 2.3 shows the contribution of ICT manufacturing to productivity growth over the 1990s, distinguishing between the first half of the decade and the second half of the decade. In most OECD countries, the contribution of ICT manufacturing to overall labour productivity growth has risen over the 1990s. This can partly be attributed to more rapid technological progress in the production of certain ICT goods, such as semi-conductors, which has contributed to more rapid price declines and thus to higher growth in real volumes (Jorgenson, 2001). However, there is a large variation in the

Table 2.1. **The impact of ICT investment on GDP growth – results from national studies**

Country	GDP growth 1990-95	GDP growth 1995-2000	Labour prod. growth 1990-95	Labour prod. growth 1995-2000	Contribution of ICT 1990-95	Contribution of ICT 1995-2000	Notes
United States							
Oliner and Sichel (2002)	–	–	1.5	2.3	0.5	1.0	1991-95; 1996-2001
Jorgenson, et al. (2002)	2.5	4.0	1.4	2.7	0.5	1.0	1990-95; 1995-99
BLS (2002)	–	–	1.5	2.7	0.4	0.9	1990-95; 1995-2000
Japan							
Miyagawa, et al. (2002)	–	–	2.2	1.4	0.1	0.4	1990-95; 1995-98
Motohashi (2002)	1.7	1.5	–	–	0.2	0.5	1990-95; 1995-2000
Germany							
RWI and Gordon (2002)	2.2	2.5	2.6	2.1	0.4	0.5	1990-95; 1995-2000
France							
Cette, et al. (2002)	0.5	2.2	1.6	1.1	0.2	0.3	1990-95; 1995-2000
United Kingdom							
Oulton (2001)	1.4	3.1	3.0	1.5	0.4	0.6	1989-94; 1994-98
Canada							
Armstrong, et al. (2002)	1.5	4.9	–	–	0.4	0.7	1988-95; 1995-2000
Khan and Santos (2002)	1.9	4.8	–	–	0.3	0.5	1991-95; 1996-2000
Australia							
Parham, et al. (2001)	–	–	2.1	3.7	0.7	1.3	89/90-94/95; 94/95-99/00
Simon and Wardrop (2001)	1.8	4.9	2.2	4.2	0.9	1.3	1991-95; 1996-2000
Gretton, et al. (2002)	–	–	2.2	3.5	0.6	1.1	89/90-94/95; 94/95-99/00
Belgium							
Kegels, et al. (2002)	1.5	2.8	1.9	1.9	0.3	0.5	1991-95; 1995-2000
Finland							
Jalava and Pohjola (2002)	–	–	3.9	3.5	0.6	0.5	1990-95; 1996-99
Korea							
Kim (2002)	7.5	5.0	–	–	1.4	1.2	1991-95; 1996-2000
Netherlands							
Van der Wiel (2002)	–	–	1.3	1.5	0.4	0.6	1991-95; 1996-2000

Source: See references.

types of ICT goods that are being produced in different OECD countries. Some countries only produce peripheral equipment, which is characterised by much slower technological progress and consequently by much less change in prices.[5]

ICT manufacturing made the largest contributions to aggregate productivity growth in Finland, Ireland, Japan, Korea, Sweden and the United States. In Finland, Ireland and Korea, close to 1 percentage point of aggregate productivity growth in the 1995-2001 period is due to ICT manufacturing.[6] The ICT-producing services sector (telecommunications and computer services)

Figure 2.3. **The contribution of ICT manufacturing to aggregate productivity growth**

Contribution to annual average labour productivity growth, percentage points

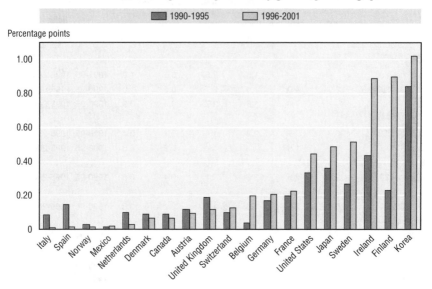

Note: 1991-95 for Germany; 1992-95 for France and Italy; 1993-95 for Korea; 1996-98 for Sweden, 1996-99 for Korea and Spain; 1996-2000 for Belgium, France, Germany, Ireland, Japan, Mexico, Norway and Switzerland.
Source: Pilat et al., (2002) and OECD STAN database.

plays a smaller role in aggregate productivity growth, but has also been characterised by rapid progress (Figure 2.4). Partly, this is linked to the liberalisation of telecommunications markets and the high speed of technological change in this market. The contribution of this sector to overall productivity growth increased in several countries over the 1990s, notably in Canada, Finland, France, Germany and the Netherlands. Some of the growth in ICT-producing services is due to the emergence of the computer services industry, which has accompanied the diffusion of ICT in OECD countries. The development of these services has been important in implementing ICT, as the firms in these sectors offer key advisory and training services and also help develop appropriate software to be used in combination with the ICT hardware.

The ICT sector is thus only an important driver of the acceleration in productivity growth in a limited number of OECD countries, notably Finland, Ireland, Japan, Korea, Sweden and the United States. This is because only few OECD countries are specialised in those parts of ICT sector that are characterised by very rapid technological progress, e.g. the production of semi-conductors and electronic computers. Indeed, much of the production of ICT

Figure 2.4. **The contribution of ICT-producing services
to aggregate productivity growth**

Contribution to annual average labour productivity growth, percentage points

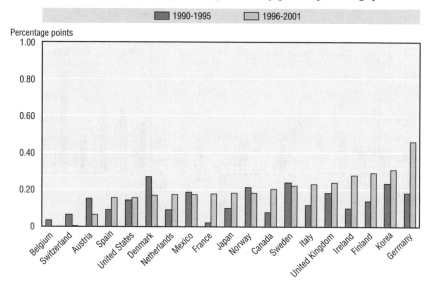

Note: See Figure 2.3 for period coverage.
Source: Pilat *et al.* (2002) and OECD STAN database.

hardware is highly concentrated, because of its large economies of scale and high entry costs. Establishing a new semi-conductor plant cost some USD 100 million in the early 1980s, but as much as USD 1.2 billion in 1999 (United States Council of Economic Advisors, 2001). In other words, a hardware sector cannot simply be set up, and only a few countries will have the necessary comparative advantages to succeed in it. In addition, a large part of the benefits of ICT production has accrued to importing countries and other users, due to terms-of-trade effects and an increased consumer surplus (Bayoumi and Haacker, 2002).

A much larger part of the economy uses ICT in the production process. Indeed, several studies have distinguished an ICT-using sector, composed of industries that are intensive users of ICT (McGuckin and Stiroh, 2001; Pilat *et al.*, 2002). Examining the performance of these sectors over time and with sectors of the economy that do not use ICT can help point to the role of ICT in aggregate performance.[7] Chapter 1 showed that services sectors such as finance and business services are intensive users of ICT.[8] Figure 2.5 shows the contribution of the key ICT-using services (wholesale and retail trade, finance, insurance and business services) to aggregate productivity growth over the 1990s.

Figure 2.5. **The contribution of ICT-using services to aggregate productivity growth**

Contribution to annual average labour productivity growth, percentage points

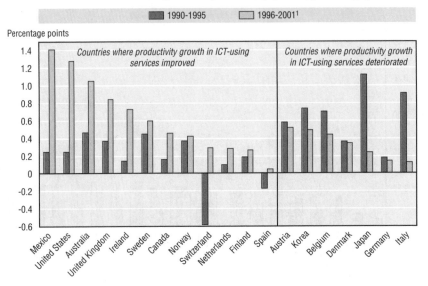

Note: See Figure 2.3 for period coverage. Estimates for Australia refer to 1996-2001.
1. Or latest available year.
Source: Pilat *et al.* (2002) and OECD STAN database.

The graph suggests small improvements in the contribution of ICT-using services in Finland, the Netherlands, Norway and Sweden, and substantial increases in Australia, Canada, Ireland, Mexico, the United Kingdom and United States. The strong increase in the United States is due to more rapid productivity growth in wholesale and retail trade, and in financial services (securities), and is confirmed by several other studies (*e.g.* McKinsey, 2001; Triplett and Bosworth, 2002). The strong increase in productivity growth in Australia has also been confirmed by other studies (Parham, 2001; Gretton *et al.*, 2002). In some countries, ICT-using services made a negative contribution to aggregate productivity growth. This is particularly the case in Switzerland in the first half of the 1990s, resulting from poor productivity growth in the banking sector.[9]

Stronger growth in labour productivity in ICT-producing and ICT-using industries could simply be due to greater use of capital. Estimates of MFP growth adjust for growth in capital stock and can help show whether ICT-using sectors have indeed improved their overall efficiency in the use of capital and labour. Breaking aggregate MFP growth down in its sectoral contributions can also help show whether changes in MFP growth should be attributed to ICT manufacturing, to ICT-using sectors, or to other sectors. Figure 2.6 shows the

Figure 2.6. **Contributions of key sectors to aggregate MFP growth,**
1990-95 and 1996-2001[1]

Contributions to annual average growth rates, in percentage points

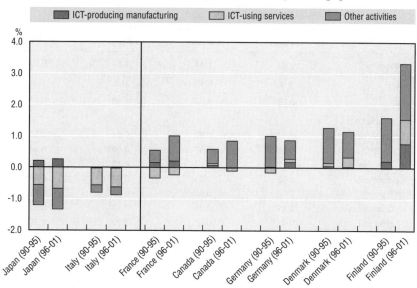

Note: Estimates are based on official estimates of capital stock and sector-specific labour shares (adjusted for labour income from self-employment). No adjustment is made for capital services.
1. Or latest available year, *i.e.* 2000 for Germany, France and Finland, 1999 for Italy, and 1998 for Japan.
Source: Pilat, Lee and Van Ark (2002) and OECD STAN database.

contribution of all activities to aggregate MFP growth for the seven countries for which estimates of capital stock at the industry level are currently available in the OECD STAN database.

Figure 2.6 shows that ICT manufacturing provided an important contribution to the acceleration in productivity growth in Finland. For ICT-using services, the MFP estimates point to growing contributions to aggregate productivity in Denmark and Finland. In several other countries, MFP growth in the ICT-using services was negative over the 1990s.

The OECD STAN database does not yet include capital stock for the United States, which implies that MFP estimates for the United States can not be derived from this source. Several studies provide estimates of the sectoral contributions to US MFP growth, however (Table 2.2). The results show considerable variation. Oliner and Sichel (2002) found no contribution of non-ICT producing industries to MFP growth; Gordon (2002) and Jorgenson, Ho and Stiroh (2002) found a relatively small contribution, while Baily (2002) and the US Council of Economic Advisors (2001) found a much more substantial contribution.[10]

Table 2.2. **Accounting for the acceleration in US productivity growth, non-farm business sector**

1995-2000 minus 1973-1995 (percentage points per year)

	Oliner-Sichel (2002), 1974-1990 *versus* 1996-2001	Gordon (2002), 1972-95 *versus* 1995-2000	US Council of Economic Advisors (2001)	Jorgenson, Ho and Stiroh (2002)
Output per hour	0.89	1.44	1.39	0.92
Cycle	n.a.	0.40	n.a.	n.a.
Trend	0.89	1.04	1.39	0.92
Contributions from:				
Capital services	0.40	0.37	0.44	0.52
IT capital	0.56	0.60	0.59	0.44
Other capital	−0.17	−0.23	−0.15	0.08
Labour quality	0.03	0.01	0.04	−0.06
MFP growth	0.46	0.52	0.91	0.47
Computer sector	0.47	0.30	0.18	0.27
Other MFP	−0.01	0.22	0.72	0.20

Source: Gordon (2002); Jorgensen, Ho and Stiroh (2002); Oliber and Sichel (2002) updated from estimates received from Dan Sichel; Council of Economic Advisors (2001) as updated in Baily (2002).

The problem with some of the studies presented in Table 2.2 (*e.g.* Oliner and Sichel, 2002 and Gordon, 2002) is that all non-ICT producing sectors are combined, and the contribution of the non-ICT producing sector to aggregate MFP growth is calculated as a residual. More detailed examination for the United States suggests that this residual is indeed small, but typically made up of a positive contribution from wholesale and retail trade and financial services to MFP growth, and a negative contribution of other service sectors. A recent study by Triplett and Bosworth (2002) finds a relatively strong pick-up in MFP growth in certain parts of the US service sector. They estimated that MFP growth in wholesale trade accelerated from 1.1% annually to 2.4% annually from 1987-1995 to 1995-2000. In retail trade, the jump was from 0.4% annually to 3.0%, and in securities the acceleration was from 2.9% to 11.2%. Combined with the relatively large weight of these sectors in the economy, this translates into a considerable contribution to more rapid aggregate MFP growth of these ICT-using services.

There is therefore evidence of strong MFP growth in the United States in ICT-using services. More detailed studies suggest how these productivity changes due to ICT use in the United States could be interpreted. First, a considerable part of the pick-up in productivity growth can be attributed to retail trade, where firms such as WalMart used innovative practices, such as the appropriate use of ICT, to gain market share from its competitors (McKinsey, 2001). The larger market share for WalMart and other productive firms raised average productivity and also forced WalMart's competitors to improve their own performance. Among the other ICT-using services, securities accounts also for a large part of the pick-up in productivity growth in the 1990s. Its strong performance has been attributed to a combination of buoyant financial markets (*i.e.* large trading volumes), effective use of ICT (mainly in automating trading processes) and stronger competition (McKinsey, 2001; Baily, 2002). These impacts of ICT on MFP are therefore primarily due to efficient use of labour and capital linked to the use of ICT in the production process. They are not necessarily due to network effects, where one firms' use of ICT has positive spillovers on the economy as a whole.

Spillover effects may also play a role, however, as ICT investment started earlier, and was stronger, in the United States than in most OECD countries (Colecchia and Schreyer, 2001; Van Ark *et al.*, 2002a). Moreover, previous OECD work has pointed out that the US economy might be able to achieve greater benefits from ICT since it got its fundamentals right before many other OECD countries (OECD, 2001a). Indeed, the United States may have benefited first from ICT investment ahead of other OECD countries, as it already had a high level of competition in the 1980s, which it strengthened through regulatory reforms in the 1980s and 1990s. For example, early and far-reaching liberalisation of the telecommunications sector boosted competition in dynamic segments of the ICT market. The combination of sound macroeconomic policies, well functioning institutions and markets, and a competitive economic environment may thus be at the core of the US success. A recent study by Gust and Marquez (2002) confirms these results and attributes relatively low investment in ICT in European countries partly to restrictive labour and product market regulations that have prevented firms from getting sufficient returns from their investment.

The United States is not the only country where ICT use may already have had impacts on MFP growth. Studies for Australia (Parham *et al.*, 2001; Simon and Wardrop, 2001; Gretton *et al.*, 2002) suggest that a range of structural reforms have been important in driving the strong uptake of ICT by firms and have enabled these investments to be used in ways that generate productivity gains. This is particularly evident in wholesale and retail trade and in financial intermediation, where most of the Australian productivity gains in the second half of the 1990s have occurred.

ICT and competitive effects

Part of the contribution of ICT-producing industries to growth and productivity can be attributed to competitive effects, *i.e.* the entry of new firms. Previous OECD work has shown that aggregate productivity growth can be broken down in several components, a part due to growth within existing

Figure 2.7. **Estimated industry entry rates relative to the total business sector**

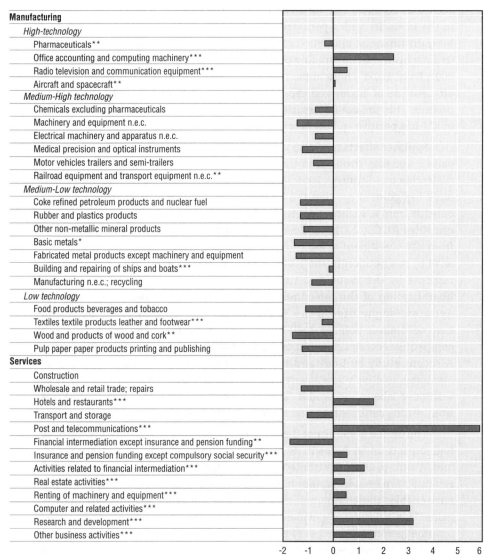

* Indicates significance at 10%; ** at 5%; *** at 1%.

Note: The figures reported in the graph are the industry fixed-effects in an entry equation that includes country, size and time-fixed effects.

Source: OECD (2003b).

firms, a part due to the entry of new firms and the exit of declining firms, and a part due to more productive firms gaining market shares. The importance of these components differs considerably between industries (Scarpetta *et al.*, 2002). In ICT-related manufacturing industries, the entry of new firms is particularly important, more so than in other manufacturing sectors (Figure 2.7). ICT-related services industries, such as post and telecommunications and computer services, also tend to have high entry rates, confirming that new firms play an important role in industries characterised by rapid technological change.

The dynamic character of the ICT-producing sector (and of other high-technology industries) is also visible in the rapid employment expansion of new firms in these industries (Figure 2.8). New entrants in these sectors – once they survive – grew much more rapidly than firms in other parts of the economy. Firms in the United States typically grow more rapidly than firms in other OECD countries, however (Scarpetta *et al.*, 2002). Figures 2.7 and 2.8 both confirm that newly created firms have provided an important contribution to growth in the ICT sector.

The OECD work on firm turnover also pointed to important differences between Europe and the United States. Compared with the European Union, the United States is characterised by: i) a smaller (relative to the industry average) size of entering firms; ii) a lower labour productivity level of entrants relative to the average incumbent; and iii) a much stronger (employment) expansion of successful entrants in the initial years which enable them to reach a higher average size. These differences in firms' performance can only partly be explained by statistical factors or differences in the business cycle (see Bartelsman *et al.*, 2002 for more details), and seem to indicate a greater degree of experimentation amongst entering firms in the United States. US firms take higher risks in adopting new technology and opt for potentially higher results, whereas European firms take fewer risks and opt for more predictable outcomes. This is likely related to differences in the business environment between the two regions; the US business environment permits greater experimentation as barriers to entry and exit are relatively low, in contrast to many European countries. The next chapter returns to this issue, as it proves to be of relevance in the adoption of ICT.

The impact of ICT on aggregate productivity growth

The evidence presented above shows that ICT has had considerable impacts on productivity growth in the second half of the 1990s, and into 2001. These effects are threefold:

1. In several countries with strong growth performance, notably Australia, Canada and the United States, investment in ICT capital has supported

Figure 2.8. **Net employment gains amongst surviving firms in high-tech industries, 1990**

Net employment gains as a percentage of initial employment

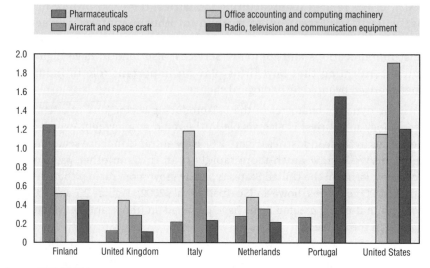

Source: OECD (2003b).

labour productivity growth. The available evidence suggests that these impacts have not disappeared with the slowdown, as ICT investment is slowly recovering.

2. In a number of countries, notably Finland, Ireland and Korea, ICT production has provided an important contribution to aggregate labour and multi-factor productivity growth.

3. In a number of OECD countries, notably Australia and the United States, there is evidence that sectors that have invested heavily in ICT, notably service sectors such as distribution and financial services, have been able to achieve more rapid MFP growth. This link between ICT use and MFP growth is also visible at the aggregate level; countries that have invested most in ICT in the 1990s have often also seen the largest increase in MFP growth over the 1990s (Figure 2.9).

A key question is the extent to which these effects are still visible in aggregate productivity growth now the economies of many OECD countries have slowed down and as parts of the ICT sector have entered a down-turn. While aggregate trends in productivity are influenced by a range of factors (OECD, 2001a), ICT is commonly considered to have contributed to a structural improvement in certain OECD countries, notably those where ICT is

Figure 2.9. **Pick-up in MFP growth and increase in ICT investment**

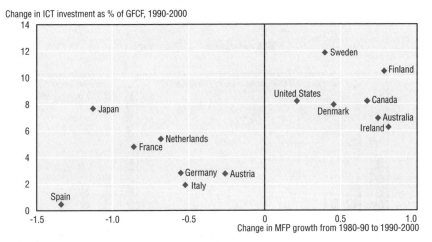

Change in ICT investment as % of GFCF, 1990-2000

Correlation coefficient = 0.66; T-statistic = 3.03.

Source: ICT investment from sources quoted in Figure 1.1, MFP growth from OECD (2003b).

considered to be important, *e.g.* Australia and the United States. Figure 2.10 shows the available evidence for key OECD countries as regards changes in labour productivity, measured as GDP per hour worked.

The top panel of Figure 2.10 shows that a limited number of OECD countries substantially increased labour productivity growth from the 1980s to the 1990s, including Australia, Canada, Denmark, Ireland, Sweden and the United States. Some of the large OECD economies, notably France, Germany, Italy and Japan experienced a substantially decline in labour productivity growth over this period. The bottom panel of the graph shows the developments in labour productivity growth over the first half of the 1990s compared with the second half. The results indicate that labour productivity growth in Australia, Ireland and the United States continued to improve over the 1990s. In several other OECD countries, cyclical conditions improved from the first to the second half of the 1990s and some countries, including France and Germany, experienced more rapid labour productivity growth over this period.[11]

A more detailed look at the data for the most recent years shows that in most OECD countries, labour productivity growth dropped sharply in 2001, as output growth in the OECD area slowed down. In some countries, such as Australia and Ireland, however, labour productivity growth continued to be strong in 2001. In both Canada and the United States, labour productivity growth picked up in 2002 after a dip in 2001 (Figure 2.11).

Figure 2.10. **Changes in GDP per hour worked in OECD countries, 1980-2001**[1]

a) Annual compound growth rates, in percentages, 1980-90 versus 1990-2001

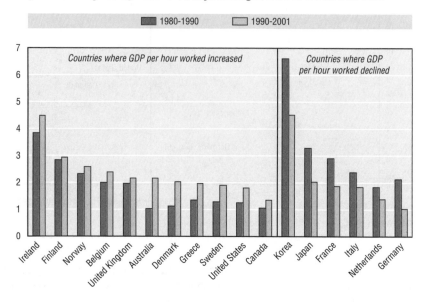

b) Annual compound growth rates, in percentages, 1990-95 versus 1995-2001

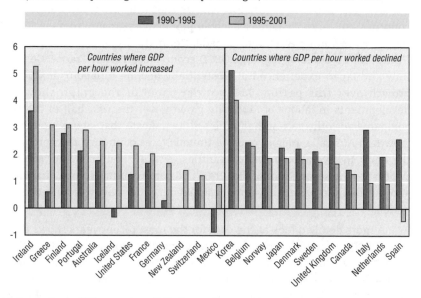

1. Estimates for the 1980s are only available for 18 OECD countries, reflecting the availability of data on hours worked.

Source: OECD; GDP and employment based on OECD Economic Outlook database, hours worked from OECD Employment Outlook.

Figure 2.11. **Recent trends in labour productivity growth,**
United States and Canada

Business sector, output per hour, percentage change from previous year

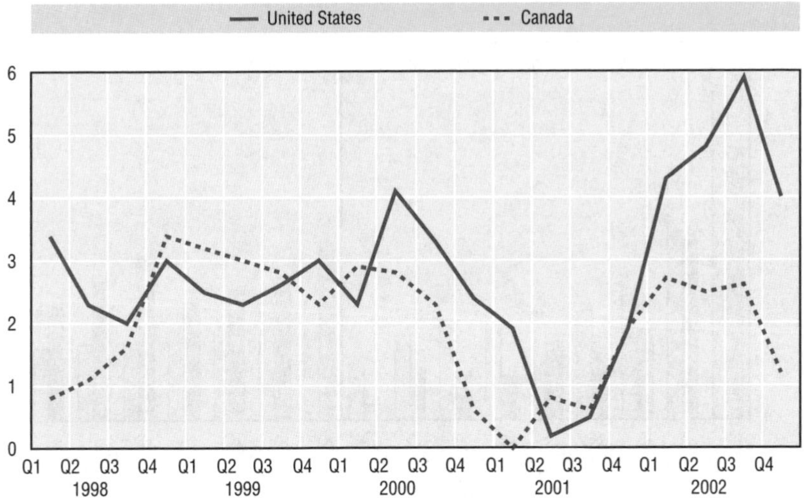

──── United States   ▪▪▪ Canada

Source: United States from Bureau of Labour Statistics, "Productivity and Costs – Fourth Quarter and Annual Averages 2002", 6 March 2003, *www.bls.gov*; Canada from Statistics Canada, "Labour Productivity, Hourly Compensation and Unit Labour Costs – Annual and Fourth Quarter 2002", *The Daily*, 14 March 2003, *www.statcan.ca*

Estimates of multifactor productivity growth are not available for many countries for the most recent years, as estimates of capital services are typically only available with some delay. Recent estimates of trends in MFP growth for a broad range of OECD countries are shown in Figure 2.12. These estimates confirm the evidence in previous OECD work, *i.e.* a structural improvement in MFP growth from the 1980s to the 1990s in Australia, Canada, Ireland, the Nordic countries, New Zealand and the United States.

The aggregate productivity trends therefore continue to point to a structural improvement in productivity growth in certain OECD countries, *e.g.* Australia, Canada and the United States, all countries that are among the key examples of ICT-led growth. This suggests that the impacts of ICT on productivity could continue in the years to come and that ICT remains a key factor for overall growth performance.

Figure 2.12. **Trends in MFP growth, 1980-90 *versus* 1990-2000**

Business sector, percentage change at annual rate

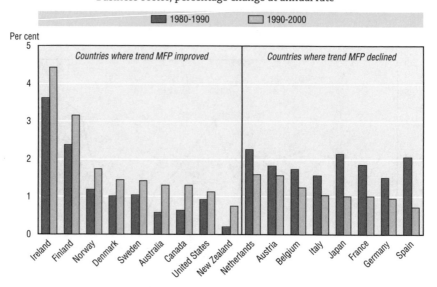

Source: OECD (2003b), *The Sources of Economic Growth in OECD Countries*, Paris. See Scarpetta *et al.* (2000) for methodological details.

Notes

1. The estimates in Figure 2.2 differ from those released in prior OECD work (notably Colecchia and Schreyer, 2001) due to data revisions in OECD countries, updates to the series, the change from estimates for the business sector to those for the economy as a whole, and some minor methodological changes that are discussed in Schreyer *et al.* (2003).

2. Although not necessarily to more rapid growth for the economy as a whole. See Schreyer (2001).

3. An international roadmap for the production of semiconductors is published by the International Technology Roadmap for Semi-conductors (ITRS). See *http://public.itrs.net/*

4. US computer prices have declined less since 1999 than they did between 1995 and 1999, which seems partly linked to reduced competitive pressure in the production of computer chips (McKinsey, 2001).

5. The large product variety also affects productivity comparisons. Some countries, such as the United States, use hedonic price indexes to capture rapid quality changes in the ICT-producing sector. This typically raises productivity growth for these sectors compared to countries that do not use these methods. However, the US hedonic price index can not simply be used (or adapted) for other countries, as the quality changes that are implicit in the US price index for ICT manufacturing may not be appropriate for a country producing only computer terminals or peripheral equipment. See Pilat *et al.*, 2002, for details.

6. Data for 2001 show a sharp slowdown in ICT production in Finland, and consequently a decline in the contribution of this sector to aggregate productivity growth.

7. A more systematic method would be to examine the link between ICT use and productivity performance by industry. However, estimates of ICT capital by industry are currently only available for some countries.

8. Certain manufacturing industries, *e.g.* printing and publishing, are also intensive users of ICT. These are not distinguished here, as the discussion has typically focused more on ICT use in services. Van Ark *et al.* (2002b) provides evidence on productivity growth in ICT-using manufacturing.

9. Poor measurement of productivity in financial services may be partly to blame. The OECD is currently working with member countries to improve methods to capture productivity growth in this sector.

10. The differences between the various US studies are partly due to the data sources and methodology used, as well as the timing of various studies.

11. The estimates shown here are not adjusted for the business cycle. Previous OECD work suggested that the conclusions did not change much when trend-adjusted estimates of productivity growth were used in the analysis (see Scarpetta *et al.*, 2000).

ISBN 92-64-10128-4
ICT and Economic Growth: Evidence from OECD Countries,
Industries and Firms
© OECD 2003

Chapter 3

ICT and firm-level performance

Abstract.   This chapter provides evidence on the contribution of ICT use to business performance, based on detailed firm-level studies. It demonstrates that the use of ICT indeed contributes to improved business performance, but only when it is complemented by other investments and actions at the firm level, such as changes in the organisation of work and changes in workers skills. The chapter also shows that ICT is no panacea; investment in ICT does not compensate for poor management, lack of skills, lack of competition, or a low ability to innovate. Not all firms will therefore succeed in generating returns from their ICT investments; many will fail. In addition, drawing the benefits from ICT investment takes time. Moreover, there are important cross-country differences in firms' use of ICT. Firms in the United States are characterised by a much higher degree of experimentation in their use of ICT than European firms; they take higher risks and opt for potentially higher outcomes.

Does ICT use matter?

The previous chapter has shown that ICT investment contributed to growth in most OECD countries in the 1990s, and that ICT production contributed to growth in some OECD countries. It also showed that ICT-using industries in the United States and Australia experienced a strong increase in productivity growth in the second half of the 1990s, partly due to their use of ICT. Few other countries have thus far experienced similar productivity gains in ICT-using services, although some aggregate evidence also suggests that the growth in MFP may be linked to the productivity-enhancing benefits from the use of ICT. Nevertheless, much of the current interest in ICT is linked to the potential economic benefits arising from its use in the production process. If the rise in MFP due to ICT were only a reflection of rapid technological progress in the production of computers, semi-conductors and related products and services, there would not be effects of ICT use on MFP in countries that are not already producers of ICT. For ICT to have benefits on MFP in countries that do not produce ICT goods, the use of ICT would need to be beneficial too.

The strongest evidence for the impact of ICT use comes from firm-level evidence, however. ICT use may have several impacts at this level. For example, the effective use of ICT may help firms gain market share at the cost of less productive firms, which could raise overall productivity. In addition, the use of ICT may help firms expand their product range, customise the services offered, or respond better to client demand; in short, to innovate. Moreover, ICT may help reduce inefficiency in the use of capital and labour, e.g. by reducing inventories. These effects might all lead to higher productivity growth. These, and related, effects have long been difficult to capture in empirical studies, contributing to the so-called "productivity paradox". However, a growing number of firm-level studies provide evidence on such impacts, suggesting that the productivity paradox has largely been solved (Box 3.1).

The diffusion of ICT may also have impacts that go beyond individual firms as it may help establish ICT networks, which produce greater benefits (the so-called spillover effects) the more customers or firms are connected to the network. For example, the spread of ICT may reduce transaction costs, which can lead to a more efficient matching of supply and demand, and enable the growth of new markets that were not feasible before. Increased use of ICT may also lead to greater scope and efficiency in the creation of knowledge, which can lead to an increase in productivity (Bartelsman and

Box 3.1. **The productivity paradox – has it been solved?**

The Solow paradox, attributed to economist Robert Solow, who once observed that computers are everywhere except in the productivity data, was appropriate during much of the 1980s and early 1990s, when the rapid diffusion of computing technology seemed to have little impact on productivity growth (Solow, 1987). Many studies in the 1970s and 1980s showed negative or zero impacts of investment in ICT on productivity. Many of these focused on labour productivity, which made the findings surprising as investment in ICT adds to the productive capital stock and should thus, in principle, contribute to labour productivity growth. Later studies did find some evidence of a positive impact of ICT on labour productivity. Some also found evidence that ICT capital had larger impacts on labour productivity than other types of capital, suggesting that there might be spillovers from ICT investment.

Studies over the past decade have pointed to several factors that contributed to the productivity paradox. First, some of the benefits of ICT were not picked up in the productivity statistics (Triplett, 1999). This is mainly a problem in the service sector, where most ICT investment occurs. For instance, the improved convenience of financial services due to automated teller machines (ATMs) is only counted as an improvement in the quality of financial services in some OECD countries. Similar problems exist for other activities such as insurance, business services and health services. ICT may have aggravated the problems of measuring productivity, as it allows greater customisation, differentiation and innovation in the services provided, most of which is difficult to capture in statistical surveys. Progress towards improved measurement has been made in some sectors and some OECD countries, but this remains an important problem in examining the impact of ICT on performance.

A second reason is that the benefits of ICT use might have taken a considerable time to emerge, as did the impacts of other key technologies, such as electricity. The diffusion of new technologies is often slow and firms can take a long time to adjust to them, e.g. in changing organisational arrangements, upgrading the workforce or inventing and implementing effective business processes. Moreover, assuming ICT raises MFP in part via the networks it provides; it takes time to build networks that are sufficiently large to have an effect on the economy. ICT has diffused very rapidly in many OECD countries in the 1990s and many recent empirical studies find a larger impact of ICT on economic performance than studies that were carried out with data for the 1970s or 1980s.

Box 3.1. **The productivity paradox – has it been solved?** (cont.)

A third reason is that many early studies that attempted to capture the impact of ICT at the firm level were based on relatively small samples of firms, drawn from private sources. If the initial impact of ICT on performance was small, such studies might find little evidence, as it would easily get lost in the econometric "noise". It is also possible that such samples were not representative of the total population. Moreover, several studies have suggested that the impact of ICT on economic performance may differ between activities, implying that a sectoral distinction in the analysis is important. More recent studies based on large samples of (official) data and covering several industries are therefore more likely to find an impact of ICT than earlier studies. In addition, early studies used a wide variation of data on ICT and ICT diffusion, often of unknown quality. Much progress has been made in recent years in measuring ICT investment and the diffusion of ICT technologies, implying that the range of available data is broader, more robust and statistically sounder than previous data.

Hinloopen, 2002). These spillover effects would drive a wedge between the impacts of ICT that can be observed at the firm level and those at the sectoral or aggregate level. Combining these three levels of information, as is done in this study, helps to shed light on this issue. This chapter examines the evidence on the role of ICT use on the basis of firm-level evidence. The next section summarises evidence on the impact of ICT on firm performance, whereas the third section examines the factors that affect the adoption of ICT and the size of ICT's impact on firm performance. Section four explores some evidence on cross-country differences at the firm level, whereas a final section looks at the link between firm-level evidence and more aggregate data.

The impacts of ICT at the firm level

A number of studies have summarised the early literature on ICT, productivity and firm performance (e.g. Brynjolfsson and Yang, 1996). Many of these studies found no, or a negative, impact of ICT on productivity. Most of these studies also primarily focused on labour productivity and the return to computer use, not on MFP or other impacts of ICT on business performance. Moreover, most of these studies used private sources, since official sources were not yet available (see Annex II). Recent work by statistical offices, using official data, has provided many new insights in the role of ICT. To help guide this work with firm-level data, OECD worked closely with an expert group, composed of researchers and statisticians from 13 OECD countries (Box 3.2). This group worked with the OECD Secretariat to generate further evidence on

> **Box 3.2. Participants in the OECD firm-level project on ICT and business performance**
>
> This study was conducted by the OECD Directorate for Science, Technology and Industry in co-operation with experts from 13 member countries. The main contacts were:
>
> **Australia:** Dean Parham and Paul Gretton (Productivity Commission), Sheridan Roberts (Australian Bureau of Statistics).
>
> **Canada:** John Baldwin (Statistics Canada).
>
> **Denmark:** Peter Bøegh Nielsen (Statistics Denmark).
>
> **Finland:** Petri Rouvinen (ETLA – Research Institute of the Finnish Economy) and Mika Maliranta (ETLA and Statistics Finland).
>
> **France:** Thomas Heckel (INSEE).
>
> **Germany:** Thomas Hempell (ZEW Centre for European Economic Research).
>
> **Italy:** Fabiola Riccardini, Carlo DeGregorio and Alessandro Zeli (ISTAT), Carlo Milana (ISAE – Institute of Studies for Economic Planning).
>
> **Japan:** Kazuyuki Motohashi (Research Institute of Economy, Trade and Industry and Hitosubashi University).
>
> **Netherlands:** Eric Bartelsman (Free University of Amsterdam), George van Leeuwen and Henry van der Wiel (CPB Netherlands Bureau of Economic Policy Analysis).
>
> **Sweden:** Anders Wiberg (IPTS – Swedish Institute for Growth Policy Studies) and Anders Hintze (Statistics Sweden).
>
> **Switzerland:** Maja Huber (Swiss Federal Statistical Office) and Heinz Hollenstein (KOF – Institute for Business Cycle Research).
>
> **United Kingdom:** Tony Clayton and Kathryn Waldron (Office for National Statistics), Jonathan Haskel and Chiara Criscuolo (University of London).
>
> **United States:** B.K. Atrostic and Ron Jarmin (Center for Economic Studies, US Bureau of the Census) and Patricia Buckley (US Department of Commerce).

the link between ICT and business performance. Their work and that of other researchers is reported in the remainder of this chapter.

The use of ICT and advanced technologies is positively linked to firm performance

There is evidence from many firm-level studies, and for many OECD countries, that ICT use has a positive impact on firm performance. These impacts can vary. Figure 3.1 illustrates a typical finding from many firm-level

Figure 3.1. **Relative productivity of advanced technology users and non-users**

Manufacturing sector in Canada, 1988 *versus* 1997

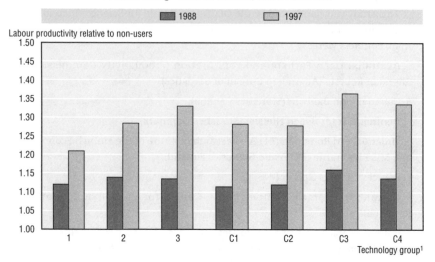

Note: The following technology groups are distinguished: Group 1 (software); Group 2 (hardware); Group 3 (communications); Group C1 (software and hardware); Group C2 (software and communications); Group C3 (hardware and communications); Group C4 (software, hardware and communications).
1. The graph shows the relative productivity on technology users compared to groups not using any advanced technology.
Source: Baldwin and Sabourin (2002).

studies that ICT-using firms have better productivity performance. It shows that Canadian firms that used either one or more ICT technologies had a higher level of productivity than firms that did not use these technologies. Moreover, the gap between technology-using firms and other firms increased between 1988 and 1997, as technology-using firms increased relative productivity compared to non-users. The graph also suggests that some ICT technologies are more important in enhancing productivity than other technologies; communication network technologies being particularly important. Figure 3.2 illustrates another important finding, based on findings from Australian firm-level data (Gretton *et al.*, 2002). It shows that the use of computers has a positive effect on MFP growth in the mid-1990s, *i.e.* before the peak in ICT investment, with considerable variation across industries.

The evidence shown in Figures 3.1 and 3.2 is confirmed by many other studies (Box 3.3) that also point to other impacts of ICT on economic performance. For example, firms using ICT typically pay higher wages. In addition, the studies show that the use of ICT does not guarantee success; many of the firms that improved performance thanks to their use of ICT were

Figure 3.2. **Estimated contribution of ICT to multifactor productivity growth in Australia**

1994-95 to 1997-98, in percentage points

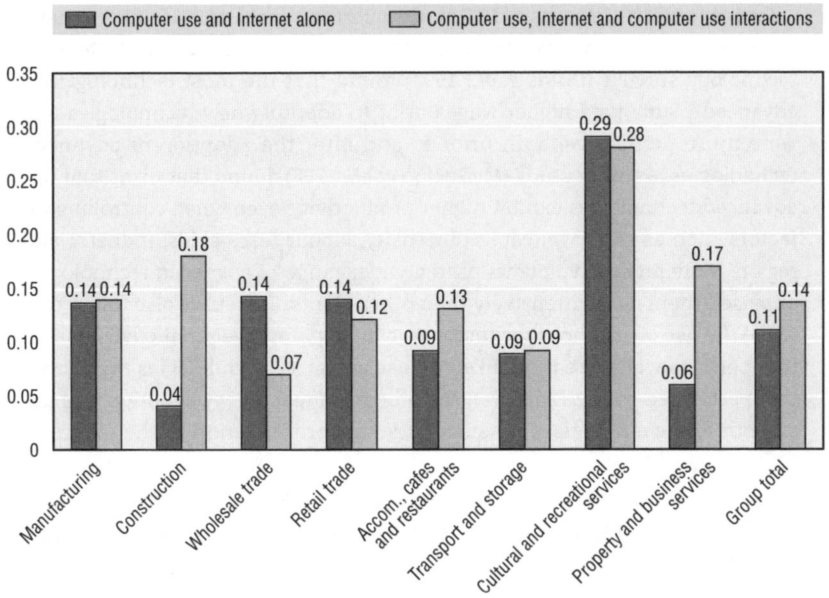

already experiencing better performance than the average firm. Moreover, the benefits of ICT appear to depend on sector-specific effects and are not found in equal measure in all sectors.

ICT can help firms to gain market share

There is also evidence that ICT can help firms in the competitive process. For the United States, Doms *et al.* (1995) found that increases in the capital intensity of the product mix and in the use of advanced manufacturing technologies were positively correlated with plant expansion and negatively with plant exit. For Canada, Baldwin *et al.* (1995a) found that establishments using advanced technologies gained market share at the expense of non-users. Technology users also enjoyed a significant labour productivity advantage over non-users, except for establishments that only used fabrication and assembly technologies. Relative labour productivity grew fastest in establishments using inspection and communications technologies and in those able to combine and integrate technologies across the different stages of the production process. Technology users were also able to offer higher wages than non-users.

Source: Gretton *et al.* (2002).

Box 3.3. **The impacts of ICT on firm performance – selected studies**

Evidence on the impacts of ICT on firm performance can be found in many studies. For example, a study for the United States using data from technology surveys (Doms et al., 1997) found that the most technologically advanced plants paid higher wages prior to adopting new technologies and were more productive, both prior to and after the adoption of advanced technologies. Another study (McGuckin et al., 1998), found that firms that use advanced technologies exhibit higher productivity, even when controlling for factors such as size, age, capital intensity, labour force skills, industry and region. More productive plants used a wider range of advanced technologies and used them more intensively than other plants. The study also found that while the use of advanced technologies can help improve productivity, plants that perform well were more likely to use advanced technologies than those that performed poorly. Moreover, the process of technology adoption was not smooth and characterised by substantial experimentation in the choice of technology.

More recent studies for the United States focused more on ICT and broadly confirmed these findings. Stolarick (1999a, 1999b) used IT investment data from the Annual Survey of Manufactures to explore the link between IT spending and productivity in manufacturing. Stolarick (1999a) found a positive relationship between IT spending and productivity, but one that varied between industries, and concluded that industry mix was thus an important explanatory factor driving aggregate findings. Stolarick (1999b) found that low productivity plants sometimes spend more on IT than high productivity plants, in an effort to compensate for poor productivity performance.

The evidence is not restricted to the United States. For France, Greenan and Mairesse (1996) matched firm data with data from surveyed employees on computer utilisation. They found very significant and positive effects linking computer use and labour productivity. In a more recent study, Crepon and Heckel (2000) used data on ICT investment at the firm and found that the effects of computer diffusion on growth were concentrated in a number of industries. In a more recent study, Biscoup et al. (2002) used firm-level data to focus on the impact of the decline in computer prices on marginal costs and the demand for labour and skills. They found that the fall in the price of computers is associated with an upward shift in the demand for skilled workers and a negative shift in the demand for unskilled workers. This effect was specific to computers and could not be found for other types of capital. The study found a high elasticity for computers, which may be due to inputs such as organisational change or embodied technology.

Box 3.3. **The impacts of ICT on firm performance – selected studies** *(cont.)*

For Italy, Milana and Zeli (2001) investigated how ICT affects production performance and technical efficiency. They found a correlation between ICT and technical efficiency in the majority of industrial sectors considered. In general, positive correlations were found in all four groups of industries defined on the basis of R&D intensity of production. De Gregorio (2002) found that micro-enterprises, *i.e.* enterprises with fewer than 10 employees, that had high ICT endowments tended to be characterised by higher innovation, R&D, training, strong inter-enterprise relations, higher productivity and higher earnings. De Panizza *et al.* (2002) found that ICT adoption in Italy was associated with previous performance of the firm.

A larger number of studies have been carried out for Canada. Baldwin and Diverty (1995) linked panel data from the Census of Manufacturers to data from a technology survey. They found that plant size and plant growth were closely related to the incidence and intensity of technology use, an indication that technology use is closely linked to the "success" of a plant. Another study (Baldwin *et al.* 1999) provided an intertemporal perspective on the use of these advanced technologies on the basis of the 1989, 1993 and 1998 technology use surveys for Canada. Among its findings are that increased use of advanced communications technologies in the 1993-1998 period was linked to the facts that these plants had superior performance in the 1980s. Moreover, foreign-owned plants had higher rates of technology adoption than domestic-owned plants in 1989 with the gap widening from 1989 to 1993. In addition, industries that innovated with regards to machinery and equipment or intermediate products that are diffused to other sectors tended to make greater use of advanced technology, suggesting that these two go hand in hand.

Bartelsman *et al.* (1996) used data from a technology use survey in the Netherlands. The study found that adoption of advanced technology is associated with higher labour productivity, higher export intensity, and larger size. Firms that employed advanced technologies in 1992 had higher productivity and employment growth in the preceding period.

In a more recent study for Canada, Baldwin and Sabourin (2002) found that a considerable amount of market share was transferred from declining firms to growing firms over a decade. At the same time, the growers increased their productivity relative to the declining firms. Those technology users that were using communications technologies or that combined technologies from several different technology classes increased their relative productivity the most. In turn, gains in relative productivity were accompanied by gains in market share. Other factors that were associated with gains in market share were the presence of R&D facilities and other innovative activities.

Computer networks play a key role

Some ICT technologies may be more important to strengthen firm performance than others. Computer networks may be particularly important, as they allow a firm to outsource certain activities, to work closer with customers and suppliers, and to better integrate activities throughout the value chain (Atrostic and Gates, 2001). These technologies are often considered to be associated with network or spillover effects.

In recent years, more data have become available on this technology. For the United States, Atrostic and Nguyen (2002) were the first in linking computer network use (both EDI and Internet) to productivity. The study found that average labour productivity was higher in plants with networks and that the impact of networks was positive and significant after controlling for several production factors and plant characteristics. Networks were estimated to increase labour productivity by roughly 5%, depending on the model specification. Atrostic et al. (2002) confirmed these findings.

Similar work has been carried out for Japan. Motohashi (2001) used the Basic Survey on Business Structure and Activities, which provides information about the networks being used by the firm, certain organisational characteristics of the firm (e.g. the degree of outsourcing), and the occupational structure of the firm. He found that the impact of direct business operation networks, such as production and logistic control systems, on productivity was much clearer than that of back office supporting systems, such as human resource management and management planning systems. Firms with networks were also found to have a larger share of white-collar workers and to outsource more production activities. Atrostic et al. (2002) also provided evidence for Japan and found that both interfirm and intrafirm networks are correlated with higher MFP levels in firms. Open networks, such as the Internet, as well as EDI networks, were particularly important.

Work in Germany has also focused on computer networks. Bertschek and Fryges (2002) were one of the first studies to examine the decision to implement business-to-business (B2B) electronic commerce. They showed that skills and firm size both have a positive and significant impact on e-commerce use. International competition, as measured by exports, also affects the decision to implement B2B, as does the firm's previous use of EDI. The most significant effect is linked to networks; the more firms in an industry that already use B2B, the more likely it is that the firm will also implement B2B. For the United Kingdom, Criscuolo and Waldron (2003) examined the role of computer networks using a similar approach as Atrostic and Nguyen (2002). They found that the use of networks had an important impact on productivity growth, but primarily through electronic purchasing, not through selling. This result confirms that networks can help firms improve the management of their supply chain.

Firms in the services sector also benefit from ICT

The work with firm-level data is also broadening to the services sector, where ICT use is more widespread than in manufacturing. For example, Doms, Jarmin and Klimek (2002) constructed a new linked dataset for US retail trade, bringing together a range of different sources. The study's results show that growth in the US retail sector involved the displacement of traditional retailers by sophisticated retailers introducing new technologies and processes, thus confirming the sectoral evidence on the US distribution sector discussed in Chapter 2.

The impacts of ICT on performance in different sectors of the economy may also be linked to the specific technologies that are being used in different sectors. Figure 3.3 presents evidence for the United Kingdom, which suggests that financial intermediation is the sector most likely to use network

Figure 3.3.   **Use of ICT network technologies by activity, United Kingdom, 2000**[1]

Percentage of all firms, business weighted

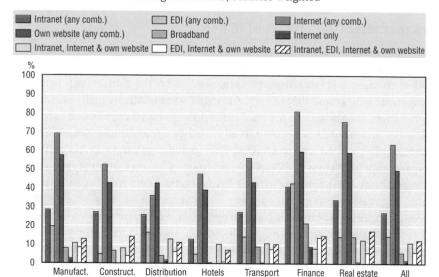

1.   Broadband includes xDSL and all other broadband connections.

Source: Clayton and Waldron (2003).

technologies, including broadband technology, and also the sector to use combinations of network technologies. The combination of several network technologies shows that these sectors are intensive users of information and thus have the greatest scope to benefit from ICT.

There is also growing evidence for other OECD countries that ICT can be beneficial to service sector performance. For Germany, Hempell (2002a) showed significant productivity effects of ICT in the German service sector. Experience gained from past process innovations helped firms to make ICT investments more productive. ICT investment may thus have contributed to growing productivity differences between firms, and potentially also between countries. A comparative study for Germany and the Netherlands (Hempell et al., 2002) confirmed the link between ICT and innovation in the German service sector, and also found such a link for the services sector of the Netherlands. Moreover, the study found that ICT capital had a significant impact on productivity in the Netherlands' services sector. In another study for the Netherlands, Broersma and McGuckin (2000) found that computer investments have a positive impact on productivity and that the impact is greater in retail than in wholesale trade. The study also found that flexible employment practices in retail trade were related to computer use. For Australia, Gretton et al. (2002) found positive impacts of ICT use on labour and MFP growth in several services sectors, in both sectoral and firm-level analysis.

Factors that affect the impact of ICT

The evidence summarised above suggests that the use of ICT does have impacts on firm performance, but primarily, or only, when accompanied by other changes and investments. Early studies on the rates of return to ICT investment suggested that the returns to ICT were relatively high compared to other investments in fixed assets. This is now commonly attributed to the fact that ICT investment is accompanied by other expenditures, which are not necessarily counted as investment. This includes expenditure on skills and organisational change. This is also confirmed by many empirical studies that suggest that ICT primarily affects firms where skills have been improved and organisational changes have been introduced. The role of these complementary factors is also raised in the literature on co-invention (Bresnahan and Greenstein, 1996), which argues that users help make investment in technologies, such as ICT, more valuable through their own experimentation and invention. Without this process of "co-invention", which often has a slower pace than technological invention, the economic impact of ICT may be limited. The firm-level evidence also suggests that the uptake and impact of ICT differs across firms, varying according to size of firm, age of the firm, activity, etc. This section looks at some of this evidence and discusses the main complementary factors for ICT investment.

ICT use is complementary to skills

A substantial number of longitudinal studies address the interaction between technology and human capital, and their joint impact on productivity performance (Bartelsman and Doms, 2000). Although few longitudinal databases include data on worker skills or occupations, some address human capital through wages, arguing that wages are positively correlated with worker skills. For the United States, Krueger (1993) used cross-sectional data and found that workers using computers were better paid than those that do not use computers. Dunne and Schmitz (1995) found that workers employed in establishments that use advanced technologies also paid higher wages. Doms et al. (1997) found no correlation between technology adoption and wages, however, and concluded that technologically advanced plants paid higher wages both before and after the adoption of new technologies. A more recent study by Luque and Miranda (2000) found that technological change in US manufacturing was skill-biased.

For France, some studies are also available. The French data include details about worker characteristics, which allow more detailed analysis. Entorf and Kramarz (1998) linked a variety of official INSEE statistics to examine the interaction between computer use and wages. They found that computer-based technologies are often used by workers with higher skills. These workers became more productive when they got more experience in using these technologies. The introduction of new technologies also contributed to a small increase in wage differentials within firms. Caroli and Van Reenen (1999) found that French plants that introduce organisational change were more likely to reduce their demand for unskilled workers than those that did not. Shortages in skilled workers might therefore reduce the probability of organisational changes. Greenan et al. (2001) also found evidence of a skill bias in the use of computers. They examined the late 1980s and early 1990s and found strong positive correlations between indicators of computerisation and research on the one hand, and productivity, average wages and the share of administrative managers on the other hand. They also found negative correlations between these indicators and the share of blue-collar workers.

For the United Kingdom, Haskel and Heden (1999) used the UK's Annual Respondents Database (ARD) together with a set of data on computerisation. They found that computerisation reduces the demand for manual workers, even when controlling for endogeneity, human capital upgrading and technological opportunities. Caroli and Van Reenen (1999) found evidence for the United Kingdom that human capital, technology and organisational change are complementary, and that organisational change reduced the demand for unskilled workers.

Studies for Canada also point to the complementarity between technology and skills. For example, Baldwin et al. (1995b) found that use of advanced technology was associated with a higher level of skill requirements. In Canadian plants using advanced technologies, this often led to a higher incidence of training. They also found that firms adopting advanced technologies increased their expenditure on education and training. A follow-up study (Baldwin et al., 1997) found that plants using advanced technologies paid higher wages to reward the higher skills required to operate these technologies. A more recent study (Sabourin, 2001) found that establishments adopting advanced technologies often reported labour shortages of scientists, engineers and technical specialists. However, the most technologically advanced establishments were often able to address these shortages.

For Australia, Gretton et al. (2002) found that the positive benefits of ICT use on MFP growth were typically linked to the level of human capital and the skill base within firms, as well as firms' experience in innovation, their application of advanced business practices and the intensity of organisational change within firms. For Germany, Falk (2001b; 2001c) found that firms with a higher diffusion of ICT employed a larger fraction of workers with a university degree as well as ICT specialists. A greater penetration of ICT was negatively related to the share of both medium and low-skilled workers.

The majority of these micro-level studies thus confirm the complementarity between technology and skills in improving productivity performance. Many also found that computers are a skill-biased technology, i.e. increasing the demand for skilled workers and reducing the demand for unskilled workers. This relationship can also be discerned at the aggregate level, although the causality is not clear; countries with a high share of highly skilled ICT workers in total occupations have had higher investment in ICT than those with fewer highly skilled ICT workers (Figure 3.4).

A few studies have also looked at other worker-related impacts. For example, Luque and Miranda (2000) found that the skill-biased technological change associated with the uptake of advanced technologies also affects worker mobility. The larger the number of advanced technologies adopted by a plant, the higher is the probability of exit of the worker. Their interpretation is that workers at technologically advanced plants have higher unobserved ability, and therefore can get a higher opportunity wage when they exit. This finding has been confirmed in other studies for both France and the United States (Entorf and Kramarz, 1997; Doms et al., 1997). The other mechanism at work is that less skilled workers tend to be pushed to plants that are less technologically advanced.

Figure 3.4. **ICT investment is associated with high skills in ICT**

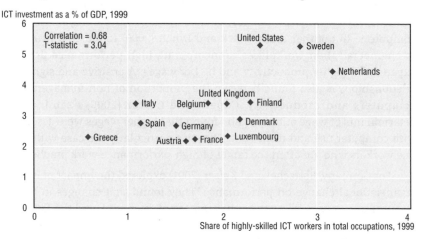

ICT investment as a % of GDP, 1999

Correlation = 0.68
T-statistic = 3.04

United States
Sweden
Netherlands
United Kingdom
Italy   Belgium   Finland
Spain   Germany   Denmark
Greece   Austria   France   Luxembourg

Share of highly-skilled ICT workers in total occupations, 1999

Source: ICT investment from sources quoted in Figure 2.1; ICT occupations from OECD (2002), *Measuring the Information Economy.*

Organisational change is key to making ICT work

Closely linked to human capital is the role of organisational change. Studies typically find that the greatest benefits from ICT are realised when ICT investment is combined with other organisational changes, such as new strategies, new business processes and practices, and new organisational structures. In the past, workers were required to perform specialised tasks within the framework of standardised production processes. In today's economy, they are often given responsibilities in different domains, for which multiple skills and the ability to work in teams are required. This phenomenon is reflected in the large variety of new work practices that are being implemented by firms. These include, *inter alia*, teamwork, flatter management structures, and employee involvement and suggestion schemes. The common element among these practices is that they entail a greater degree of responsibility of individual workers regarding the content of their work and, to some extent, a greater proximity between management and labour. Because such organisational change tends to be firm-specific, empirical studies show on average a positive return to ICT investment, but with a huge variation across organisations.

Several US studies with official statistics have addressed ICT's link to human capital and organisational change. Black and Lynch (2001) found that the implementation of human resource practices is important for productivity, *e.g.* giving employees greater voice in decision-making, profit-sharing mechanisms and new industrial relations practices. Unionised firms that adopt these practices were found to have higher productivity than firms that were not unionised, while unionised firms that did not adopt these practices had lower productivity. They also found that productivity was higher in firms with a large proportion of non-managerial employees that use computers. In another study, Black and Lynch (2000) found that firms that re-engineered their workplaces to incorporate high-performance practices experienced higher productivity and higher wages. A positive and significant relationship was also found between the proportion of non-managers using computers and productivity. Capelli and Carter (2000) examined the determinants of wage outcomes and found that higher wages were associated with computer use and teamwork. This was particularly the case with front-line workers who are often the target of high-performance work practices.

For Germany, Bertschek and Kaiser (2001) explored the impact of ICT and organisational change on performance. They found that changes in human resource practices, such as the enhancement of teamwork and the flattening of hierarchies, did not significantly affect firm's output elasticities with respect to ICT capital, non-ICT capital and labour. The study also found no evidence of significant differences in returns to scale. It did, however, find that the introduction of organisational changes raised overall labour productivity. Studies in Germany have also explored the link between ICT use, organisational change and human capital. Falk (2001a) used results from the 1995 and 1997 Mannheim Innovation Panel in Services (MIP-S). He found that the introduction of ICT and the share of training expenditures were important drivers of organisational changes, such as the introduction of total quality management, lean administration, flatter hierarchies and delegation of authority. The study found that organisational changes had a positive impact on actual employment and on expected employment, apart from unskilled groups. The prospects for such organisational changes may be affected by policy barriers, however. In a 2000 survey of German firms, more than 23% of firms outside the ICT sector cited legal restrictions as a barrier to the adoption of ICT and 19% of non-ICT firms mentioned internal resistance within the firm as a barrier to uptake (Hempell et al., 2002).

For France, Greenan and Guellec (1998) found that the use of advanced technologies and the skills of the workforce were both positively linked to organisational variables. Organisations that enabled communication within the firm and that innovated at the organisational level seemed better able to create the conditions for a successful uptake of advanced technologies. Moreover, these changes also seemed to increase the ability of firms to adjust to changing market conditions through technological innovation and the reduction of inventories.

For the United Kingdom, Caroli and Van Reenen (1999) used the Workplace Industrial Relations Survey, which included questions about the introduction of micro-electronic technology and organisational change. The study found that organisational change, technology and skills were complementary. More specifically, it found that organisational change reduced the demand for unskilled workers; that such change was retarded by increases in regional skill price differentials; and that organisational change had the largest productivity impacts in establishments with larger initial skill endowments. For the Netherlands, Broersma and McGuckin (2000) also found that computer use was linked to the introduction of flexible employment practices, *e.g.* greater use of temporary and part-time workers.

For Switzerland, Arvantis (2003) investigated the influence of ICT, new practices of workplace organisation, formal education and job-related training on firm productivity. The econometric results show that ICT and human capital are positively correlated with productivity; the effect of organisation is also positive but not statistically significant. Moreover, firms with a high share of highly skilled employees are those with a high productivity of ICT; new organisational practices did not seem to contribute to a higher productivity of either ICT or human capital.

Firm size and age affect the impact of ICT

A substantial number of studies have looked at the relationship between ICT and firm size. This relationship can go in different ways. The first question is whether there is a difference in the uptake of ICT by size classes. This question has been addressed in a large number of studies in many countries, *e.g.* Dunne (1994) for the United States, Baldwin and Diverty (1995) for Canada, or Bartelsman *et al.* (1996) for the Netherlands. Most of these find that the adoption of advanced technologies, such as ICT, increases with the size of firms and plants.

Figure 3.5 confirms this result for the United Kingdom, with recent data for a variety of network technologies used in different combinations. It shows that large firms of over 250 employees are more likely to use network technologies such as Intranet, Internet or EDI than small firms; they are also more likely to have their own Web site. However, small firms of between 10 and 49 employees are more likely to use Internet as their only ICT network technology. Large firms are also more likely to use a combination of network technologies. For example, over 38% of all large UK firms use Intranet, EDI and Internet, and also have their own Web site, as opposed to less than 5% of small firms. Moreover, almost 45% of all large firms already use broadband technologies as opposed to less than 7% of small firms.

These differences are partly due to the different uses of the network technologies by large and small firms. Large firms may use the technologies to redesign information and communication flows within the firm, and to integrate these flows throughout the production process. Some small firms

Figure 3.5.   **Use of ICT network technologies by size class, United Kingdom, 2000**

Percentage of all firms, business-weighted

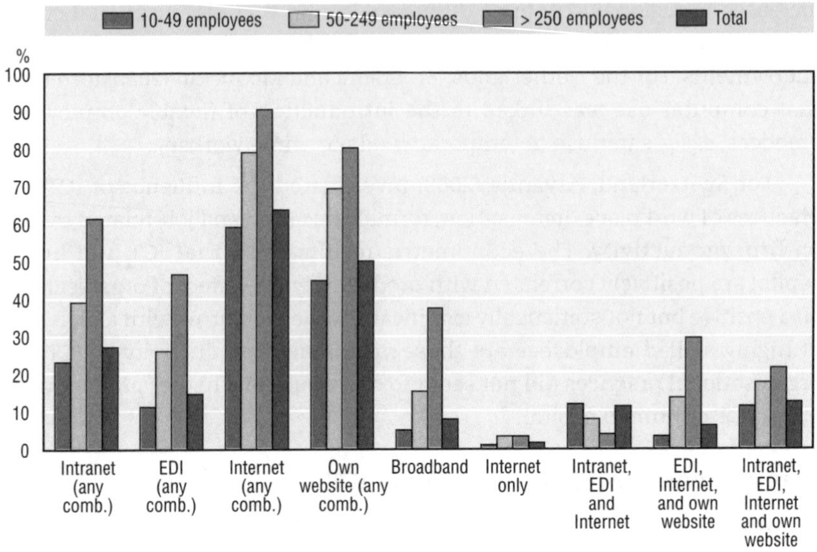

Source: Clayton and Waldron (2003).

only use the Internet for marketing purposes. Moreover, skilled managers and employees often help in making the technology work in large firms (Gretton *et al.*, 2002).

There is also a question whether ICT has an effect on the size of firms or changes the boundaries of firms over time. This question is linked to the expectation that ICT might help lower transaction costs and thus changes the functions and tasks that should be carried out within firms and those that could be carried out outside the firm boundaries. This issue has been researched by only few firm-level studies, most of which use private data. For example, Hitt (1998) found that increased use of ICT was associated with decreases in vertical integration and increased diversification. Moreover, firms that were less vertically integrated and more diversified had a higher demand for ICT capital. Motohashi (2001) found that firms with computer networks outsourced more activities.

The link between size and age is also important, as it provides a link to firm creation. Dunne (1994) found that the impact of age on the likelihood of adopting advanced technologies was quite small. Luque (2000) confirmed this result, but found that age may have a role depending on plant size. Small new plants were more likely to adopt advanced technologies than small old plants.

Ownership, competition and management are important

Firm-level studies also point to the importance of ownership changes and management in the uptake of technology. For example, a study by McGuckin and Nguyen (1995) for the food processing industry found that plants with above-average productivity were more likely to change owners and that the acquiring firms also tended to have above-average productivity. Plants that changed owners typically improved productivity following the change. Moreover, ownership changes appeared associated with the purchase or integration of advanced technologies and better practices into new firms. These results were confirmed by Baldwin (1995) in a study for the Canadian manufacturing sector.

Some studies also point to the impact of competition. A study by Baldwin and Diverty (1995) found that foreign-owned plants were more likely to adopt advanced technologies than domestic plants. For Germany, Bertschek and Fryges (2002) found that international competition was an important factor driving a firm's decision to implement B2B electronic commerce. These findings should be linked to the results of several firm-level studies that show that the implementation of advanced technologies can help firms to gain market share and may reduce the likelihood of plant exit (e.g. Doms et al., 1995; Doms, Jarmin and Klimek, 2002; Baldwin et al., 1995a; Baldwin and Sabourin, 2002).

Stolarick (1999b) found that low productivity plants may sometimes spend more on IT than high productivity plants, in an effort to compensate for their poor productivity performance. The study suggests that management skill should therefore be taken into account as an additional factor when investigating the IT productivity paradox.

ICT use is closely linked to innovation

Several studies point to an important link between the use of ICT and the ability of a company to adjust to changing demand and to innovate. The possibility of such a link is also visible in aggregate data; those countries that have invested most in ICT also have the largest share of patents in ICT (Figure 3.6).

The best example of this link is found in work on Germany by ZEW, as this draws on innovation survey results. For example, Licht and Moch (1999) draw on the Mannheim Innovation Panel in Services (MIP-S). They found that information technology has important impacts on the qualitative aspects of service innovation, but not on productivity. Hempell (2002a) also used this survey and found that firms that had introduced process innovations in the past were particularly successful in using ICT; the output elasticity of ICT capital for these firms was estimated to be about 12%, about four times that of

Figure 3.6. **ICT investment is accompanied by rapid innovation in ICT**

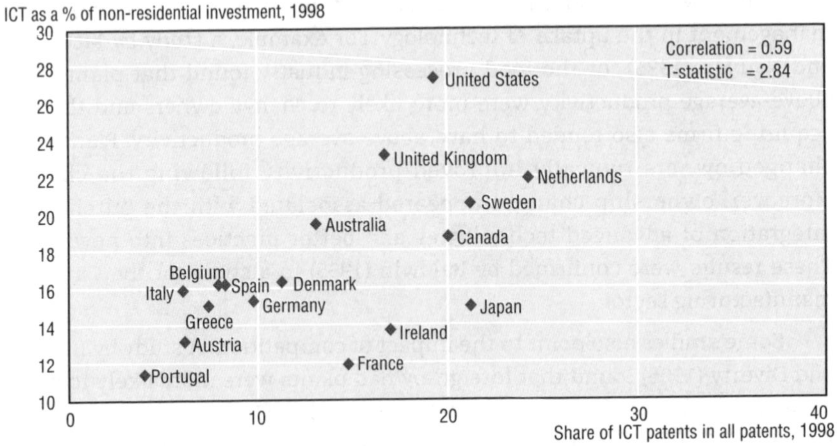

ICT as a % of non-residential investment, 1998

Correlation = 0.59
T-statistic = 2.84

United States
United Kingdom
Netherlands
Sweden
Australia
Canada
Belgium
Italy Spain Denmark
Germany
Greece Japan
Austria Ireland
France
Portugal

Share of ICT patents in all patents, 1998

Source: ICT investment from Figure 2.1; ICT patents from OECD (2002), *Measuring the Information Economy.*

other firms. This suggests that the productive use of ICT is closely linked to innovation in general, and to the re-engineering of processes in particular. Moreover, the introduction of ICT has many similarities with innovation, as it is risky and uncertain, with potentially positive outcomes.

Studies in other countries also confirm this link. For example, Greenan and Guellec (1998) found that organisational change and the uptake of advanced technologies increased the ability of firms to adjust to changing market conditions through technological innovation. For the Netherlands, Hempell *et al.* (2002) found that services firms that engaged in permanent non-technological innovation benefited more from ICT than those that did not.

The impacts of ICT use often only emerge over time

Given the time it takes to adapt to ICT, it should not be surprising that the benefits of ICT may only emerge over time. This can be seen in the relationship between the use of ICT and the year in which firms first adopted ICT. Figure 3.7 shows evidence for the United Kingdom. It shows that among the firms that had already adopted ICT in or before 1995, close to 50% bought using electronic commerce in 2000. For firms that only adopted ICT in 2000, less than 20% bought using e-commerce. The graph also suggests that firms move towards more complex forms of electronic activity over time; out of all firms starting to use ICT prior to 1995, only 3% had not yet moved beyond the straightforward use of ICT in 2000. Most had established an Internet site, or

Figure 3.7. **Relationship between the year of ICT adoption
and the current degree of e-activity**

As a percentage of all firms adopting ICT in specific year, business-weighted

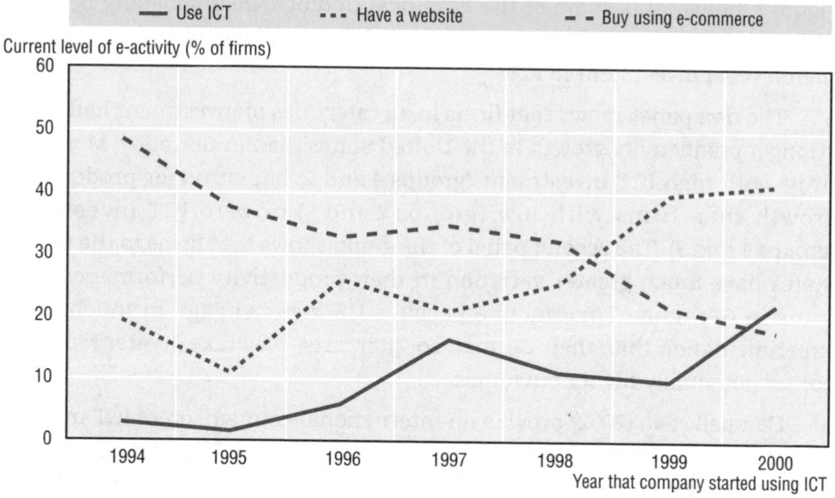

Note: The graph shows the percentage of firms engaged in a specific type of e-activity in 2000, out of all the firms starting to use ICT in that year.

Source: Clayton and Waldron (2003).

bought or sold through e-commerce. Out of the firms adopting ICT in 2000, over 20% had not yet gone beyond the simple use of ICT.

The role of time also emerges from analysis for Australia. Gretton et al., 2002 used firm level information on productivity growth and the duration of computer use to examine the dynamics of the impact of the introduction of computers. They found that computers had a positive effect on MFP growth that varied between industries (Figure 3.2). They also found that the positive effect was largest in the earlier years of uptake but appeared to taper off as firms returned to "normal" growth after the productivity boost of the new technology. This indicates that the ultimate productivity effect from adoption of ICT is a step up in levels, rather than a permanent increase in the rate of growth. However, further technical developments can set further productivity-enhancing processes in motion.

Does the impact of ICT at the firm level differ across countries?

Cross-country studies on the impact of ICT at the firm level are still relatively scarce, primarily since many of the original data sources were of an ad hoc nature and not comparable across countries. In recent years, the growing similarity of official statistics is enabling more comparative work, which has been further encouraged in the recent OECD work with statistical offices. An example of such a study is a recent comparison between the United States and Germany (Haltiwanger *et al.* 2002), that examines the relationship between labour productivity and measures of the choice of technology. Figure 3.8 illustrates some of the empirical findings, distinguishing between different categories of firms according to their total level of investment and their level of investment in ICT.

The first panel shows that firms in all categories of investment had much stronger productivity growth in the United States than in Germany. Moreover, firms with high ICT investment (groups 4 and 6) had stronger productivity growth than firms with low (groups 2 and 5) or zero ICT investment (groups 1 and 3). The second panel of the graph shows that firms in the United States have much greater variation in their productivity performance than firms in Germany. This may be because US firms engage in much more experimentation than their German counterparts; they take greater risks and opt for potentially higher outcomes.

Hempell *et al.* (2002) provide an international comparison of ICT impacts in the services sector in Germany and the Netherlands. They found that ICT capital deepening raises labour productivity in both countries and that investment in ICT capital appears to be complementary to innovation in both countries. They also found some differences, *e.g.* innovation had a more direct impact on productivity in Germany than in the Netherlands. Other international comparisons of business performance and the impact of ICT are currently underway on the basis of microdata (Atrostic *et al.*, 2002); these should contribute to further insights and help explain cross-country differences in the benefits that are being drawn from ICT.

Concluding remarks on the firm-level evidence

Examining the role of ICT at the aggregate, sectoral and firm level raises some difficult questions (see Gretton *et al.*, 2002). The firm-level evidence suggests that ICT use is beneficial – though under certain conditions – to firm performance in all countries for which micro-level studies have been conducted. However, the aggregate and sectoral evidence is less conclusive about the benefits of ICT use. It shows that investment in ICT capital has contributed to growth in most OECD countries, and that the ICT-producing sector has contributed to productivity growth in some OECD countries. There is, however, little evidence that ICT-using industries have experienced more rapid productivity growth, the United States and Australia being the major exceptions. There are several reasons why this may be the case and why aggregate evidence may differ from firm-specific evidence:

1. Aggregation across firms and industries, as well as the effects of other economic changes, may disguise some of the impacts of ICT in sectoral and

Figure 3.8. **Differences in productivity outcomes between Germany and the United States**

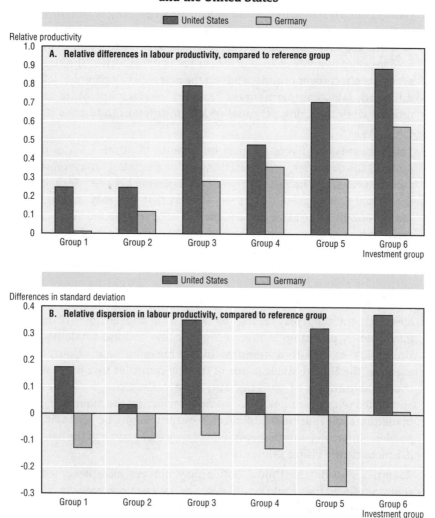

Note: Differences are in logs and are shown relative to a reference group of zero total investment and zero investment in ICT. The groups are distinguished on the basis of total investment (0, low, high) and ICT investment (0, low, high). Group 1 has low overall investment and zero ICT investment. Group 2 has low overall investment and low ICT investment. Group 3 has high overall investment and zero ICT investment. Group 4 has low overall investment and high ICT investment. Group 5 has high overall investment and low ICT investment. Group 6 has high overall investment and high ICT investment.

Source: Haltiwanger, Jarmin and Schank (2002).

aggregate analysis that are more evident from firm level analysis. This may also be because the impacts of ICT depend on other factors and policy changes, which may differ across industries. Regulatory reform in specific sectors and specific countries, financial services for example, may already have allowed ICT to strengthen performance, while lack of reform may still hold back productivity change in other sectors or countries. The size of the aggregate effects over time depends on the rate of development of ICT, their diffusion, lags, complementary changes, adjustment costs and the productivity-enhancing potential of ICT in different industries (Gretton et al., 2002).

2. The firm-level benefits of ICT may be larger in the United States than in other OECD countries, and thus show up more clearly in aggregate and sectoral evidence. The results from the comparison between Germany and the United States, presented above, suggest that this may indeed be the case. Given the more extensive diffusion of ICT in the United States, and its early start, this interpretation should not be surprising; in particular if it takes time before the benefits from ICT become apparent. Moreover, the conditions under which ICT is beneficial to firm performance, such as sufficient scope for organisational change, might be more firmly established in the United States than in other countries.

3. Measurement may play a role as well. The impacts of ICT may be insufficiently picked up in macroeconomic and sectoral data outside the United States, due to differences in the measurement of output. For example, the United States is one of the few countries that have changed the measurement of banking output to reflect the convenience of automated teller machines. Since services sectors are the main users of ICT, inadequate measurement of service output might be a considerable problem. Improvements in measurement may make some of the benefits of ICT more clearly visible (Annex I).

4. Countries outside the United States may not yet have benefited from spillover effects that could create a wedge between the impacts observed for individual firms and those at the macroeconomic level. The discussion above has already suggested that the impacts of ICT may be larger than the direct returns flowing to firms using ICT. For example, ICT may lower transaction costs, which can improve the functioning of markets (by improving the matching process), and make new markets possible. Another effect that can create a gap between firm-level returns and aggregate returns is ICT's impact on knowledge creation and innovation. ICT enables more data and information to be processed at a higher speed and can thus increase the productivity of the process of knowledge creation. A greater use of ICT may thus gradually improve the functioning of the economy.

Such spillover effects may already have shown up in the aggregate statistics in the United States, but not yet in other countries.

5. The state of competition may also play a role in the size of spillover effects. In a large and highly competitive market, such as the United States, firms using ICT may not be the largest beneficiaries of investment in ICT. Consumers may extract a large part of the benefits, in the form of lower prices, better quality, improved convenience, and so on. In other cases, firms that are upstream or downstream in the value chain from the firms using ICT might benefit from greater efficiency in other parts of the value chain. For example, Criscuolo and Waldron (2003) demonstrate large productivity impacts for firms purchasing through computer networks, not for firms selling through networks. In countries with limited competition, firms might be able to extract a greater part of the returns, and spillover effects might thus be more limited. Further cross-country research may help to address these questions, and provide new insights in the extent of ICT-related spillovers.

The empirical evidence presented in this chapter and Chapter 2 also allows a number of general conclusions to be drawn. First, while ICT investment has dropped off during the recent slowdown, it is likely to increase once the recovery gets underway. Technological progress in ICT goods and services is continuing at a rapid pace, driving prices down and contributing to a wide range of new services and applications. For example, the release of increasingly powerful microprocessors is projected to continue for the foreseeable future, which will encourage ICT investment and support further productivity growth. The level of ICT investment is likely to be lower than that observed prior to the slowdown, however, as the 1995-2000 period was characterised by some one-off investment peaks, *e.g.* investments related to Y2K and the diffusion of the Internet (Gordon, 2003). Second, further technological progress in ICT production should imply a continued positive contribution of the ICT sector to MFP growth, notably in those countries with large ICT-producing sectors. Third, productivity growth in the United States, Canada and Australia, examples of ICT-led growth, has continued to be strong during the recent slowdown, suggesting that part of the acceleration in productivity growth over the second half of the 1990s was indeed structural. Finally, ICT networks have now spread throughout much of the OECD business sector, implying that will increasingly be made to work to enhance productivity and business performance. But these effects will also depend on whether policy makers can foster a business environment that enables firms to make smart and effective use from the technology.

ANNEX I

Some implications for statistics

This study uses a range of new statistics, many of which have only existed for a short period. The use of these statistics allows some feedback on their relevance and applicability, *e.g.* as regards surveys on ICT use by businesses and surveys of electronic commerce. The work has also highlighted some remaining constraints in the analysis of ICT's impact on business performance. The statistical implications of the project are several.

First, the project has demonstrated the importance of databases that enable researchers and statisticians to link firm-level data derived from different statistical surveys. These databases are of particular importance since the impact of ICT depends on the use of a range of complementary investments and factors, such as skills, organisational change, management and competition. Examining any of these factors in isolation is of limited use. Statistical offices should therefore consider the potential for linking databases when designing new statistical surveys. Common firm identifiers and sample frames are important for this to happen and point to the need for integrated business registers. Databases that allow firms to be tracked over time are also highly relevant. They provide crucial insights in firm creation and destruction, and in the changing fortunes of firms, as well as the factors which influence firm dynamics.

Second, data constraints still limit the scope for comparative analysis. Many of the current sources are not yet sufficiently comparable to allow large-scale cross-country comparisons. However, more limited comparative studies are possible and provide important insights in cross-country differences. Moreover, data on certain key variables, such as organisational factors and skills, are currently only available for a limited number of countries, making international studies quite difficult. Not all variables that are of potential interest are available from statistical surveys; many variables are not available for the same periods; and combining several sources typically limits the sample of firms.

Third, the work has shown that ICT's impact remains difficult to capture at the aggregate and sectoral level in many services sectors, due to (well known) problems in the measurement of output. Further statistical work in these areas (*e.g.* financial services, business services and health) should be a priority. Firm-level studies demonstrate that productivity growth in services is positively affected by ICT, which appears to contradict the sectoral evidence in many countries. Work is now underway in many countries to address this area through the development of better deflators and output measures.

Finally, certain areas offer scope for further progress. For example, surveys that focus on the specific use of different ICT technologies by businesses, *e.g.* as regards the technologies being used and the businesses processes being affected, may offer greater prospects for examining the impact of ICT than more general surveys on ICT and Internet use. Most firms already use ICT and the Internet, and this is no longer a characteristic that helps to distinguish well performing from poorly performing firms. The relationship between ICT and the ability to innovate is potentially also of great interest, but has not yet been explored in many studies. Linking innovation surveys and ICT surveys to production surveys can provide important insight into the long-term benefits of ICT. Furthermore, surveys that provide qualitative evidence on the perceived impacts of ICT at the firm level, and that are currently under development in some OECD countries, may be a helpful complement to existing surveys and data collections as they may avoid some of the measurement problems noted above. Finally, studies in some countries have successfully linked employer and employee data to explore the role of skills and worker characteristics. Much has been achieved in recent years, but data linking and longitudinal analysis offer a great potential for future statistical analysis, with potentially important policy implications.

ANNEX II

Data for firm-level studies

Most of the early work with firm-level data on ICT and business performance, and a considerable amount of work by academic researchers, is based on private data for a sample of firms. For example, Brynjolfsson and Hitt (1997) examined more than 600 large US firms over the 1987-94 period, partly drawing on the Compustat database, while Bresnahan, Brynjolfsson and Hitt (2002) examined over 300 large US firms from the Fortune 1000. Similar studies with private data exist for other countries. Work with these private sources has helped generate interest in this area of analysis and has given an important impetus to the development of official statistics on ICT. However, private sources suffer from a number of methodological drawbacks. First, the representativeness of the sample of firms is often not known, which may imply that the results of such studies are biased. Second, the quality of the data is not always known, since the data do not necessarily confirm with accepted statistical conventions and definitions.

Over the past decade, the analysis in this area has benefited from the establishment of longitudinal databases in statistical offices. These databases cover much larger and statistically representative samples than private data, which is important given the enormous heterogeneity in plant and/or firm performance (Bartelsman and Doms, 2000). These data allow firms to be tracked over time and can be linked to many surveys and data sources. Among the first of these databases was the Longitudinal Research Database of the Center of Economic Studies (CES) at the US Bureau of the Census (McGuckin and Pascoe, 1988). Since then, several other countries have also established longitudinal databases and centres for analytical studies with these data. Examples include Australia, Canada, Finland, France, the Netherlands and the United Kingdom. The data integrated in these longitudinal databases differ somewhat between countries, since the underlying sources are not the same. However, many of the basic elements of these databases are common. The basic source for productivity-related analysis in most countries is typically a production survey or census, such as the US Annual Survey of Manufactures. These data typically cover the manufacturing sector, although longitudinal databases increasingly cover the service sector as well.

In recent years, statistical offices and researchers working with these data have increasingly become involved in analytical work on ICT's contribution to economic performance and productivity. The first of these studies typically used technology use surveys, such as the Survey of Manufacturing Technology in the Netherlands or the United States, and the Survey of Advanced Technology in Canada (see Vickery and Northcott, 1995, for an overview of these surveys). Other studies used data on IT investment derived from production or investment surveys. In recent years, more data on ICT have become available, *e.g.* from surveys of ICT use and e-commerce undertaken in many OECD countries. Moreover, innovation surveys, such as the Community Innovation Survey, often include some survey questions on computer use that can, in principle, be used for empirical analysis. Finally, several countries have other special surveys that touch on aspects of ICT use by firms. In principle, such surveys can also be used for more detailed analysis. More detail is available on the specific sources in each country is shown in Table A2.1.

Table A2.1. **Key databases for firm-level statistical analysis on ICT and business performance**

	Longitudinal databases	Production, employment	Technology use surveys	Innovation surveys	Research and Development	Other sources on ICT use	Electronic Commerce	Other surveys used
Australia	Business longitudinal survey, 1994-95 – 1997-98	Economic activity survey				Survey on business use of ICT, 1997-2001		
Canada	Longitudinal manufactures research file; Longitudinal employment analysis programme	Annual survey of manufactures	Survey of innovation and advanced technology 1993; 1998 SAT	1996/1999 survey of innovation				
Denmark		1995-99 enterprise statistics; 1995-1999 account statistics				1998 survey on enterprise use of ICT		Database for labour market research (IDA)
Finland	Longitudinal data on plants in manufacturing	Annual industrial statistics, business register on plants		Community innovation surveys	R&D surveys		E-commerce surveys	Employee statistics
France		Annual survey of enterprises (EAE); BRN employer file; SUSE firm accounts		Community innovation surveys (CIS)	R&D surveys	Survey Organisational change and informatisation (COI)		Annual social declarations (DADS); Employment structure survey
Germany (ZEW)		Service sector business survey		Mannheim innovation panel		Survey on ICT use and skill shortages	ZEW CATI survey for 2000	
Italy		Survey on economic and financial accounts			R&D assets from enterprise account	ICT assets and investment from enterprise accounts	E-commerce survey	

Table A2.1. **Key databases for firm-level statistical analysis on ICT and business performance** (cont.)

	Longitudinal databases	Production, employment	Technology use surveys	Innovation surveys	Research and Development	Other sources on ICT use	Electronic Commerce	Other surveys used
Japan		Basic survey on business structure and activities			R&D surveys	Basic survey on business structure and activities (BSBSA)	E-commerce survey	Survey of IT workplaces (SITW)
Netherlands	Microlab data CBS (CEREM)	Production survey, 1978-1999	Automation survey, 1987-97	CIS 1994-96, 1996-98, 2000	Quadrennial, from 1985 onwards	Investment in fixed assets, 1993-99		
Switzerland			Advanced manufacturing survey, 1996	Innovation survey, every 3rd year since 1990			2002 survey on ICT and e-commerce	2000 survey on ICT and organisational change
United Kingdom	ONS business data bank/ Annual respondents databank (ARD)	Annual business inquiry 1998-2001		CIS 1997 and 2000	R&D surveys 2000-2001	ICT capital expenditure, e-commerce and software spending from ABI supplement	E-commerce surveys 2000-2001	Workplace industrial relations survey
United States	Longitudinal research database, Longitudinal business database	Census/survey of manufactures; SSEL economic census	Survey of manufacturing technology, 1988, 1991, 1993		National science foundation R&D surveys	Computer spending from ASM; Business expenditure survey, computer network use from ASM supplement		EQW national employer survey

ISBN 92-64-10128-4
ICT and Economic Growth: Evidence from OECD Countries,
Industries and Firms
© OECD 2003

Chapter 4

Policy implications

Abstract. This chapter draws implications from the empirical evidence presented in previous chapters for policy makers. It argues that governments should reduce unnecessary costs and regulatory burdens on firms to create a business environment that promotes productive investment. This involves policies to enable organisational change, to strengthen education and training systems, to encourage good management practices, and to foster innovation, e.g. in new applications, that can accompany the uptake of ICT. Moreover, policy should foster market conditions that reward the successful adoption of ICT; a competitive environment is key for this to happen. Governments will also need to work with business and consumers to shape a regulatory framework that strengthens confidence and trust in the use of ICT, notably electronic commerce. Policies to foster growth in services are important too, as ICT offers a new potential for growth in the service sector, providing that regulations that stifle change are adjusted or removed. Finally, the report reaffirms the importance of economic and social fundamentals as the key to lasting improvements in economic performance. A short set of conclusions completes the report.

The OECD Growth Study provided a number of recommendations on policies to seize the benefits of ICT and foster economic growth (OECD, 2001a). This included, *inter alia*, policies to increase competition in telecommunications, to enhance skills and encourage labour mobility, to reduce obstacles to workplace changes, and to build confidence in the use of ICT. The Growth Study also concluded that ICT is not the only factor explaining growth disparities, and that policies to bolster ICT will not on their own steer countries on to a higher growth path. Strengthening growth performance will require a comprehensive and co-ordinated set of actions to create the right conditions for future change and innovation, including policies to strengthen fundamentals, to foster innovation, to invest in human capital and to stimulate firm creation. The present study reaffirms these conclusions and provides further evidence on the appropriate policies to seize the benefits of ICT.[1]

Strengthening competition in ICT goods and services

The evidence presented in previous chapters provides new insights in the factors that influence a firm's decision to invest in ICT. Firms will decide to invest in ICT if they can make smart and effective use of their investment, and expect sufficient returns. These returns are determined by several factors, some of which can be influenced by policy makers. One important factor is the cost of the investment in ICT itself. The available evidence suggests that differences in the costs of the technology continue to play a role in determining investment patterns. Estimates of relative price levels of ICT investment for 1999 still showed a considerable variation across the OECD area (OECD, 2002c). This variation is considerably smaller than in 1993 or 1996 (OECD, 2001a), but still suggests that the United States had among the lowest price levels for ICT equipment and software. Barriers to trade, in particular non-tariff barriers related to standards, import licensing and government procurement, may partly explain the cost differentials. The higher price levels in other OECD countries may also be associated with a lack of competition within countries. In time, however, international trade and competition should further erode these cross-country price differences. Policy could help to accelerate this trend, by implementing a more active competition policy and measures to promote market openness, both domestically and internationally.

The investment and diffusion of ICT do not just depend on the cost of the investment goods themselves, but also on the associated costs of communication and use once the hardware is linked to a network. Increased competition in the telecommunications industry, thanks to extensive regulatory reform, has been of particular importance in driving down these costs. Liberalisation, and the competition it has generated, has brought tremendous benefits to OECD countries and users. Prices have declined, and continue to do so in certain market segments. Technological diffusion and new service development have been rapid, and continue to grow. Incumbent telecommunication carriers have adjusted to the new market conditions by increasing efficiency and improving levels of service. A large number of new entrants have entered the market, and while some have failed, the number of market players in many OECD countries remains large. But efforts to improve competitive conditions have not yet been sufficient. The OECD growth report called on countries to increase competition and continue with regulatory reform in the telecommunications industry to enhance the uptake of ICT. The evidence presented in this report shows that this recommendation will continue to be important in the years to come (OECD, 2003a).

Fostering a business environment for ICT adoption

A competitive environment is more likely to lead a firm to invest in ICT, as a way to strengthen performance and survive, than a more sheltered environment. Moreover, the state of competition influences firms' decisions to implement ICT applications, such as electronic commerce. Many firms do not engage in e-commerce because the market is considered too small, or because their products are not considered suitable for electronic commerce. In other cases, electronic commerce is seen as a rival to existing business models. These concerns can be genuine, but may also reflect a conservative attitude. Existing firms may wish to retain their current business model and avoid the risks associated with new investments and new business models. Start-up firms can help instil greater dynamism, introduce new business models, and invigorate mature industries. Policies to enhance firm creation are key in such markets.

The previous chapters showed that ICT is technology that has the potential to transform firms. They can use it in smart ways to improve performance, but not all firms will succeed in making the necessary changes that are needed to make the technology work. Competition and creative destruction are key in selecting the successful firms and in making them flourish and grow. If firms that are able to make ICT work succeed and grow, the benefits for the economy as a whole are greater than if poorly performing firms survive. Chapters 2 and 3 demonstrated cross-country differences in experimentation that may be important for firms wishing to gain benefits from their investments in ICT (Bartelsman *et al.* 2002). Allowing room for such experimentation is important. New firms in the United States seem to experiment more with business models than those in other OECD countries; they start at a smaller scale than European firms, but grow much more quickly when successful. This may be linked to less aversion to risk in the United States, linked to its financial system, which provides greater opportunities for risky financing to innovative entrepreneurs. Moreover, low regulatory burdens enable US firms to start at a small scale, experiment, test the market and their business model, and, if successful, expand rapidly. Moreover, if they do not succeed, the costs of failure are relatively limited. In contrast, firms in many other OECD countries are faced with high entry and exit costs. In a period of rapid technological change, greater scope for experimentation may enable new ideas and innovation to emerge more rapidly, leading to faster technology diffusion.

Investment in ICT relies not on only competition and the cost of ICT, but also on the complementary investments that need to be made by firms to draw the benefits from ICT, *e.g.* in changing the organisation of functions and tasks, or in training staff. These complementary investments are often much more costly than the initial outlays for ICT investment goods. Brynjolfsson and Hitt (2000), for example, suggest that USD 1 of ICT investment may be associated with USD 9 of investment in intangible assets, such as skills and organisational practices. Adapting the organisation of functions and tasks to ICT can be particularly costly to firms, as it often meets with resistance within the firm, and may be limited by legal constraints. Social partners and government can work together to ensure that a virtuous circle of human resource upgrading, organisational change, ICT and productivity is set in motion. This depends on workers being given a sufficient "voice" in the firm. A closer contact between management and employees can help build a high-skill, high-trust enterprise climate that facilitates change. This may also require ensuring that working time legislation and employment regulations do not hamper such change, and that collective bargaining institutions are adapted to the new environment.

Matching the skills of workers to the new technology also requires considerable investment. For ICT to be developed and used effectively, and network externalities to materialise, the right skills and competencies must be in place. Having a good supply of qualified personnel helps, but education policies, important as they are, need to be supplemented with actions to foster lifelong learning. Policies aimed at enhancing basic literacy in ICT, at building high-level ICT skills, at lifelong learning in ICT, and at enhancing the managerial and networking skills needed for the effective use of ICT, are particularly relevant. Moreover, a certain degree of labour mobility is needed to seize the new opportunities associated with ICT, which may require changes to regulations in some countries.

Another implication relates to management. Firm-level studies typically find that firms that get most out of their investment in ICT are those that firms that were already performing well in terms of gains in productivity and market shares. These firms improved performance by investing in ICT, by innovating and by adapting their organisation and workforce.[2] In contrast, many firms that invested much in ICT received no returns at all, as they were attempting to compensate for poor overall performance. This reinforces the view that ICT is no panacea, and also points to a role for management. While governments cannot directly influence management decisions, it can help create framework conditions for good management. Frameworks for good corporate governance also play a role in this context.

Policies to seize the benefits of ICT rely on fundamental economic and social stability to succeed. All of the policy areas discussed in this paper are interlinked and depend on each other. But those countries that have managed to seize the benefits of ICT were able to do so because they had been getting their fundamentals right. They owed their economic success to sound macroeconomic policies, well-functioning institutions and markets, and an orientation to build a more open and competitive economic environment. Studies for Australia, one of the key examples of ICT-driven growth, emphasise the interaction between structural reform and the uptake of ICT (Parham et al., 2001). By contrast, in those countries whose growth performances appeared to lag, some of the fundamentals were perhaps missing or were at best so weak as to make it difficult to harness the new dynamism, such as not having the right institutional set-up for new business creation.

Boosting security and trust

Businesses, governments, consumers and key infrastructures increasingly rely on the use of information networks, which are often interconnected at the global level. This raises new issues for security as these electronic networks need to be stable and ready for safe, secure and reliable use under all conditions. Legal uncertainties (uncertainty over payments, contracts, terms of delivery and guarantees) remain a barrier to electronic commerce. Likewise, business-to-consumer transactions are hampered by concerns about security of payments, opportunities for redress, and the privacy of personal data. For all users, whether businesses or consumers, the security and reliability of systems and information networks is important.

Much work is currently underway to address these concerns. Authentication and certification mechanisms are being developed to help identify users and safeguard business transactions. To counter computer viruses, hacking and other threats, OECD has drawn up new and comprehensive security guidelines that are currently awaiting implementation by OECD countries. These guidelines aim to promote a "culture of security" in the operation of information systems and networks.

With the growth of business-to-consumer e-commerce transactions, consumer complaints regarding the online environment are growing. The OECD privacy and consumer protection guidelines are an important step towards an international consensus on core protections. Continued efforts to implement these guidelines are key and will require that governments, business and civil society work together. Further exploitation of information technologies can enhance consumer trust, by facilitating access to information and improving the ability of users to protect themselves, *e.g.* through privacy enhancing technologies. But for any trust-related tool or measure to have a positive impact on trust, consumers and users must be aware of, and understand the protections afforded. Education and awareness-raising policies are therefore important. Moreover, ensuring that current laws and regulations are effectively enforced in cross-border situations is a major challenge.

Some of the slowness to do business (personal or otherwise) via the Internet is to do with attitudes. Governments can help to change these by using ICT applications themselves. Tendering public services, providing digital public services, collecting taxes or procuring goods and services online can help increase government efficiency and enhance access to public services, while having the additional benefit of building public confidence and strengthening demand.

Unleashing growth in the services sector

ICT has already brought many benefits and has the potential to bring more. The service sector is particularly important in drawing the benefits from ICT, since industries such as wholesale and retail trade, financial and business services are among the most important investors in ICT. It is in these "old economy" sectors, not in the ICT sector, that the long-term impacts of ICT use may be most important. Evidence for the United States and Australia shows that ICT has already enabled productivity growth in some of these industries.

Policies must take better account of the needs and characteristics of the services sector if they are to promote growth (OECD, 2001c). Competition and business dynamism in many services sectors remains limited due to regulatory burdens, reducing pressures to strengthen performance. Moreover, sector-specific regulations may reduce the development of new ICT applications and may limit the capability of firms to seize the benefits of ICT.

Further reform of regulatory structures is needed to promote competition and innovation and to reduce barriers and administrative rules for new entrants and start-ups. International competition is also important, and will require the reduction of trade and foreign investment barriers in services. Firm-level studies show that foreign-owned firms are often the first to adopt new technologies.

Harnessing the potential of innovation and technology diffusion

ICT is closely linked to the ability of firms to innovate, i.e. introduce new products and services, new business processes, and new applications. Firms that have already innovated achieve much better results from ICT than those that have never innovated. Moreover, ICT has helped facilitate the innovation process, for example in speeding up scientific discovery. ICT has also fostered networking, which has enabled greater outsourcing of R&D and enabled informal learning between firms, which is key to innovation in services firms. Policies to harness the potential of innovation, as outlined in the OECD Growth Study, are thus important in seizing the benefits of ICT. Moreover, such policies help foster the kind of innovative environment in which new growth opportunities will flourish. To strengthen innovation, policy needs to give greater priority to fundamental research, improve the effectiveness of public R&D funding and promote the flow of knowledge between science and industry.

Concluding remarks

Despite the slowdown in the economy and parts of the ICT sector, ICT has emerged over the past decade as a key technology with the potential to transform economic and social activity. It has already led to more rapid growth in countries where appropriate policies to reap the benefits from ICT have been put in place. Moreover, continued technological change should bring many more benefits in the future. All OECD governments can do more to exploit this technology, by fostering a business environment that encourages its diffusion and use and by building confidence and trust. However, policies to bolster ICT will not on their own lead to stronger economic performance. Indeed, economic performance is not the result of a single policy or institutional arrangement, but a comprehensive and co-ordinated set of actions to create the right conditions for future change and innovation. Policies to strengthen economic and social fundamentals are of great importance in drawing the benefits from ICT. The policy implications arising from this report thus reaffirm and elaborate those of the OECD growth report.

Policymakers have to be prepared to invest time and political capital in meeting these challenges. Many of the countries that already reaped the benefits from ICT in the 1990s reaped the fruits of their earlier efforts, *e.g.* in liberalising the telecommunications industry or in improving their business environment. Policy action will also require further examination of a range of thorny, yet unresolved issues. There is a major knowledge gap regarding which impact, if any, ICT has on the functioning of markets, including digital markets, *e.g.* in reducing transaction costs and changing the respective market power of different parties. A better understanding of ICT's impact on innovation, as well as society's ability to deal with ICT will also be essential.

Further examination of the impacts of ICT on economic performance will also require appropriate statistics. This report has benefited from new data, which have contributed to many new insights. Further work with firm-level data may be particularly important, as such studies demonstrate that ICT is no panacea, but depends on a range of other changes in the way firms go about their business.

Notes

1. Other implications for policies to seize the benefits from ICT emerge from other OECD work, notably the OECD 2003 Communications Outlook; work on security, trust and consumer protection; as well as a forthcoming study on the economic impacts of electronic business. The policy conclusions from that work and the work contained in this paper are summarised in a short booklet for the OECD 2003 Ministerial Council Meeting, entitled "Seizing the Benefits of ICT in a Digital Economy". This chapter primarily summarises the policy implications that can be drawn from the empirical analysis presented in this report.

2. The management literature provides extensive discussions on how firms can make ICT work in their particular environment. These issues are not discussed here, as government policy has little role in influencing these corporate processes.

References

Ahmad, N. (2003), "Measuring Investment in Software", STI Working Paper 2003/6, OECD, Paris.

Aizcorbe, A. (2002), "Why are Semiconductor Prices Falling So Fast? Industry Estimates and Implications for Productivity Measurement", Finance and Economics Discussion Series 2002-20, Federal Reserve Board, Washington DC.

Armstrong, P., T.M. Harchaoui, C. Jackson and F. Tarkhani (2002), "A Comparison of Canada – US Economic Growth in the Information Age, 1981-2000: The Importance of Investment in Information and Communication Technologies", Economic Analysis Research Paper Series, No. 001, Statistics Canada, Ottawa.

Arvantis, S. (2003), "Information Technology, Workplace Organization, Human Capital and Firm Productivity: Evidence for the Swiss Economy", *KOF-Arbeitspapiere/* Working Papers No. 70, February, Zurich.

Atrostic, B.K., P. Boegh-Nielsen and K. Motohashi (2002), "IT, Productivity and Growth in Enterprises: Evidence from New International Micro Data", paper presented at OECD Workshop on ICT and Business Performance, December.

Atrostic, B.K. and J. Gates (2001), "US Productivity and Electronic Processes in Manufacturing", CES WP-01-11, Center for Economic Studies, Washington DC.

Atrostic, B.K. and S. Nguyen (2002), "Computer Networks and US Manufacturing Plant Productivity: New Evidence from the CNUS Data", CES Working Paper 02-01, Center for Economic Studies, Washington DC.

Baily, M.N. (2002), "The New Economy: Post Mortem or Second Wind", *Journal of Economic Perspectives*, Vol. 16, No. 2, Spring 2002, pp. 3-22.

Baily, M.N., C. Hulten, and D. Campbell (1992), "Productivity Dynamics in Manufacturing Plants", *Brookings Papers on Economic Activity: Microeconomics*, pp. 187-267.

Baldwin, J. (1995), *The Dynamics of Industrial Competition – A North American Perspective*, Cambridge University Press, Cambridge.

Baldwin, J.R. and B. Diverty (1995), "Advanced Technology Use in Canadian Manufacturing Establishments", Working Paper No. 85, Microeconomics Analysis Division, Statistics Canada, Ottawa.

Baldwin, J.R., B. Diverty, and D. Sabourin (1995a), "Technology Use and Industrial Transformation: Empirical Perspective", Working Paper No. 75, Microeconomics Analysis Division, Statistics Canada, Ottawa.

Baldwin, J.R., T. Gray, and J. Johnson (1995b), "Technology Use, Training and Plant-Specific Knowledge in Manufacturing Establishments", Working Paper No. 86, Microeconomics Analysis Division, Statistics Canada, Ottawa.

Baldwin, J.R., T. Gray, and J. Johnson (1997), "Technology-Induced Wage Premia in Canadian Manufacturing Plants During the 1980s", Working Paper No. 92, Microeconomics Analysis Division, Statistics Canada, Ottawa.

Baldwin, J.R., E. Rama and D. Sabourin (1999), "Growth of Advanced Technology Use in Canadian Manufacturing During the 1990s", Working Paper No. 105, Microeconomics Analysis Division, Statistics Canada, Ottawa.

Baldwin, J.R. and D. Sabourin (2002), "Impact of the Adoption of Advanced Information and Communication Technologies on Firm Performance in the Canadian Manufacturing Sector", STI Working Paper 2002/1, OECD, Paris.

Baldwin, J.R., D. Sabourin and D. Smith (2002), "Impact of ICT Use on Firm Performance in the Canadian Food Processing Sector", Working Paper, Microeconomics Analysis Division, Statistics Canada, Ottawa, forthcoming.

Bartelsman, E.J. and M. Doms (2000), "Understanding Productivity: Lessons from Longitudinal Micro Datasets", Journal of Economic Literature, Vol. 38, September.

Bartelsman, E.J., G. van Leeuwen and H.R. Nieuwenhuijsen (1996), "Advanced Manufacturing Technology and Firm Performance in the Netherlands", Netherlands Official Statistics, Vol. 11, Autumn, pp. 40-51.

Bartelsman, E. A. Bassanini, J. Haltiwanger, R. Jarmin, S. Scarpetta and T. Schank (2002), "The Spread of ICT and Productivity Growth – Is Europe Really Lagging Behind in the New Economy?", Fondazione Rodolfo DeBenedetti, mimeo.

Bartelsman, E. and J. Hinloopen (2002), "Unleashing Animal Spirits: Investment in ICT and Economic Growth", mimeo.

Bayoumi, T. and M. Haacker (2002), "It's Not What You Make, It's How You Use It: Measuring the Welfare Benefits of the IT Revolution Across Countries", CEPR Discussion Papers No. 3555, Center for Economic Policy Research, London.

Bertschek, I. and U. Kaiser (2001), "Productivity Effects of Organizational Change: Microeconometric Evidence", ZEW Discussion Paper No. 01-32, ZEW, Mannheim.

Bertschek, I. and H. Fryges (2002), "The Adoption of Business-to-Business E-Commerce: Empirical Evidence for German Companies", ZEW Discussion Paper No. 02-05, ZEW, Mannheim.

Biscoup, P., B. Crépon, T. Heckel and N. Riedinger (2002), "How Do Firms Respond to Cheaper Computers? Microeconometric Evidence for France Based on a Production Function Approach", G2002/05, INSEE, April.

Black, S.E. and L.M. Lynch (2000), "What's Driving the New Economy: The Benefits of Workplace Innovation", NBER Working Paper Series, No. 7479, January.

Black, S.E. and L.M. Lynch (2001), "How to Compete: The Impact of Workplace Practices and Information Technology on Productivity", The Review of Economics and Statistics, August, Vol. 83, No. 3, pp. 434-445.

Bresnahan, T.F., E. Brynjolfsson, and L.M. Hitt (2002), "Information Technology, Workplace Organization and the Demand for Skilled Labor: Firm-Level Evidence", Quarterly Journal of Economics, Vol. 117, February, pp. 339-376.

Bresnahan, T.F. and S. Greenstein (1996), "Technical Progress and Co-Invention in Computing and the Use of Computers", Brookings Papers on Economic Activity: Microeconomics, pp. 1-77.

Broersma, L. and R.H. McGuckin (2000), "The Impact of Computers on Productivity in the Trade Sector: Explorations with Dutch Microdata", Research Memorandum GD-45, Groningen Growth and Development Centre, June.

Brynjolfsson, E. and L. Hitt (1997), "Computing Productivity: Are Computers Pulling Their Weight?", mimeo MIT and Wharton, *http://ccs.mit.edu/erik/cpg/*

Brynjolfsson, E. and L.M. Hitt (2000), "Beyond Computation: Information Technology, Organizational Transformation and Business Performance", *Journal of Economic Perspectives*, 14, pp. 23-48.

Brynjolfsson, E. and S. Yang (1996), "Information Technology and Productivity: A Review of the Literature", *mimeo, http://ecommerce.mit.edu/erik/*

Capelli, P. and W. Carter (2000), "Computers, Work Organization, and Wage Outcomes", NBER Working Paper 7987, National Bureau of Economic Research, Cambridge, MA, October.

Caroli, E. and J. Van Reenen (1999), "Organization, Skills and Technology: Evidence from a Panel of British and French Establishments", IFS Working Paper Series W99/23, Institute of Fiscal Studies, August.

Cette, G., J. Mairesse and Y. Kocoglu (2002), "Diffusion of ICTs and Growth of the French Economy over the Long Term, 1980-2000", *International Productivity Monitor*, No. 4, Spring, pp. 27-38.

Clayton, T. and K. Waldron (2003), "E-Commerce Adoption and Business Impact, A Progress Report", *Economic Trends*, forthcoming.

Colecchia, A. and P. Schreyer (2001), "The Impact of Information Communications Technology on Output Growth", STI Working Paper 2001/7, OECD, Paris.

Crepon, B. and T. Heckel (2000), " Informatisation en France: une évaluation à partir de données individuelles ", G2000/13, Document de travail (Working paper, in French), INSEE.

Criscuolo, C. and K. Waldron (2003), "Computer Network Use and Productivity in the United Kingdom", Centre for Research into Business Activity and Office of National Statistics, *mimeo.*

Dedrick, J. and K.L. Kraemer (2001), "The Productivity Paradox: Is it Resolved? Is There a New One? What Does It All Mean for Managers", Center for Research on Information Technology and Organizations, Irvine.

De Gregorio, C. (2002), "Micro Enterprises in Italy: Are ICTs and Opportunity for Growth and Competitiveness?", paper presented at OECD workshop on ICT and Business Performance, ISTAT, Rome, December.

De Panniza, A., L. Nascia, A. Nurra, F. Oropallo and F. Riccardini (2002), "ICT and Business Performance in Italy", paper presented at OECD workshop on ICT and Business Performance, ISTAT, Rome, December.

Devlin, A. (2003), "Explaining ICT Investment in OECD Countries", STI Working Papers, OECD, Paris, forthcoming.

Doms, M., T. Dunne, and M.J. Roberts (1995), "The Role of Technology Use in the Survival and Growth of Manufacturing Plants", *International Journal of Industrial Organization* 13, No. 4, December, pp. 523-542.

Doms, M., T. Dunne and K.R. Troske (1997), "Workers, Wages and Technology", *Quarterly Journal of Economics*, 112, No. 1, pp. 253-290.

Doms, M., R. Jarmin and S. Klimek (2002), "IT Investment and Firm Performance in US Retail Trade", CES Working Paper 02-14, Center for Economic Studies, Washington DC.

Dunne, T. (1994), "Plant Age and Technology Use in US Manufacturing Industries", *Rand Journal of Economics*, Vol. 25, No. 3, pp. 488-499.

Dunne, T. and J. Schmitz (1995), "Wages, Employment Structure and Employer Size-Wage Premia: Their Relationship to Advanced-technology Usage at US Manufacturing Establishments", *Economica*, March, pp. 89-107.

Entorf, H. and F. Kramarz (1997), "Does Unmeasured Ability Explain the Higher Wages of New Technology Workers?", *European Economics Review*, 41, pp. 1489-1509.

Entorf, H. and F. Kramarz (1998), "The Impact of New Technologies on Wages: Lessons from Matching Panels on Employees and on their Firms", *Economic Innovation and New Technology*, Vol. 5, pp. 165-197.

Falk, M. (2001a), "Organizational Change, New Information and Communication Technologies and the Demand for Labor in Services", ZEW Discussion Paper No. 01-25, ZEW, Mannheim.

Falk, M. (2001b), "Diffusion of Information Technology, Internet Use and the Demand for Heterogeneous Labor", ZEW Discussion Paper No. 01-48, ZEW, Mannheim.

Falk, M. (2001c), "The Impact of Office Machinery and Computer Capital on the Demand for Heterogeneous Labor", ZEW Discussion Paper No. 01-66, ZEW, Mannheim.

Gordon, R.J. (2002), "Technology and Economic Performance in the American Economy", *NBER Working Papers,* No. 8771, National Bureau of Economic Research, February.

Gordon, R.J. (2003), "Hi-Tech Innovation and Productivity Growth: Does Supply Create Its Own Demand", NBER Working Papers, No. 9437, National Bureau of Economic Research, January.

Greenan, N. and D. Guellec (1998), "Firm Organization, Technology and Performance: An Empirical Study", *Economics of Innovation and New Technology*, Vol. 6, No. 4, pp. 313-347.

Greenan, N. and J. Mairesse (1996), "Computers and Productivity in France: Some Evidence", NBER Working Paper 5836, Cambridge, MA.

Greenan, N., J. Mairesse and A. Topiol-Bensaid (2001), "Information Technology and Research and Development Impacts on Productivity and Skills: Looking for Correlations on French Firm Level Data", NBER Working Paper 8075, Cambridge, MA.

Gretton, P., J. Gali and D. Parham (2002), "Uptake and Impacts of ICT in the Australian Economy: Evidence from Aggregate, Sectoral and Firm Levels", paper presented at OECD Workshop on ICT and Business Performance, Productivity Commission, Canberra, December.

Gust, C. and J. Marquez (2002), "International Comparisons of Productivity Growth: The Role of Information Technology and Regulatory Practices", International Finance Discussion Papers, No. 727, Federal Reserve Board, May.

Haltiwanger, J., R. Jarmin and T. Schank (2002), "Productivity, Investment in ICT and Market Experimentation: Micro Evidence from Germany and the United States.", paper presented at OECD Workshop on ICT and Business Performance, December.

Haskel, J. and Y. Heden (1999), "Computers and the Demand for Skilled Labour: Industry- and Establishment-Level Panel Evidence for the UK", *The Economic Journal*, 109, C68-C79, March.

Hempell, T. (2002a), "Does Experience Matter? Productivity Effects of ICT in the German Service Sector", Discussion Paper No. 02-43, Centre for European Economic Research, Mannheim.

Hempell, T. (2002b), "What's Spurious, What's Real? Measuring the Productivity Impacts of ICT at the Firm-Level", Discussion Paper No. 02-42, Centre for European Economic Research, Mannheim.

Hempell, T., G. Van Leeuwen and H. Van Der Wiel (2002), "ICT, Innovation and Business Performance in Services: Evidence for Germany and the Netherlands", paper presented at OECD Workshop on ICT and Business Performance, December.

Hitt, L.M. (1998), "Information Technology and Firm Boundaries: Evidence from Panel Data", University of Pennsylvania, *mimeo*.

Hollenstein, H. (2002), "The Decision to Adopt Information and Communication Technologies (ICT): Explanation and Policy Conclusions", paper presented at OECD workshop on ICT and Business Performance, Institute for Business Cycle Research (KOF), Zurich, December.

Jorgenson, D.W. (2001), "Information Technology and the US Economy", *American Economic Review*, Vol. 91, No. 1, pp. 1-32.

Jorgenson, D.W., M.S. Ho and K.J. Stiroh (2002), "Projecting Productivity Growth: Lessons from the US Growth Resurgence", *Federal Reserve Bank of Atlanta Economic Review*, third quarter, pp. 1-13.

Kegels, C., M. Van Overbeke and W. Van Zandweghe (2002), "ICT Contribution to Economic Performance in Belgium: Preliminary Evidence", Working Paper 8-02, Federal Planning Bureau, Brussels, September.

Khan, H. and M. Santos (2002), "Contribution of ICT Use to Output and Labour: Productivity Growth in Canada", Working Papers 2002-7, Bank of Canada, Ottawa, March.

Kim, S.J. (2002), *The Digital Economy and the Role of Government: Information Technology and Economic Performance in Korea*, Program on Information Resources Policy, Harvard University, January.

Krueger, A.B. (1993), "How Computers Have Changed the Wage Structure: Evidence from Microdata, 1984-1989", *The Quarterly Journal of Economics*, February, pp. 33-60.

Licht, G. and D. Moch (1999), "Innovation and Information Technology in Services", *Canadian Journal of Economics*, Vol. 32, No. 2, April.

Luque, A. (2000), "An Option-Value Approach to Technology Adoption in US Manufacturing: Evidence from Plant-Level Data", CES WP-00-12, Center for Economic Studies, Washington, DC.

Luque, A. and J. Miranda (2000), "Technology Use and Worker Outcomes: Direct Evidence from Linked Employee-Employer Data", CES WP-00-13, Center for Economic Studies, Washington, DC.

McGuckin, R.H. and G.A. Pascoe (1988), "The Longitudinal Research Database: Status and Research Possibilities", *Survey of Current Business*, 68, November, pp. 30-37.

McGuckin, R.H. and S.V. Nguyen (1995), "On Productivity and Plant Ownership Change: New Evidence from the LRD", *Rand Journal of Economics,* 26, No. 2, pp. 257-276.

McGuckin, R.H., M. Strietweiser, and M. Doms (1998), "The Effect of Technology Use on Productivity Growth", *Economics of Innovation and New Technology,* Vol. 7, pp. 1-26.

McGuckin, R.H. and K.J. Stiroh (2001), "Do Computers Make Output Harder to Measure?", *Journal of Technology Transfer,* Vol. 26, pp. 295-321.

McGuckin, R.H. and B. Van Ark (2001), "Making the Most of the Information Age – Productivity and Structural Reform in the New Economy", *Perspectives on a Global Economy,* Research Report R-1301-01-RR, Conference Board, October.

McKinsey (2001), *US Productivity Growth 1995-2000: Understanding the Contribution of Information Technology Relative to Other Factors,* McKinsey Global Institute, Washington, DC. October.

Milana, C. and A. Zeli (2001), "The Contribution of ICT to Production Efficiency in Italy: Firm-Level Evidence using DEA and Econometric Estimations", STI Working Paper 2002/13, OECD, Paris.

Miyagawa, T., Y. Ito and N. Harada (2002), "Does the IT Revolution Contribute to Japanese Economic Growth?", JCER Discussion Paper No. 75, Japan Center for Economic Research, Tokyo.

Motohashi, K. (2001), "Economic Analysis of Information Network Use: Organisational and Productivity Impacts on Japanese Firms", Research and Statistics Department, METI, *mimeo.*

Motohashi, K. (2002), "IT Investment and Productivity Growth of the Japanese Economy and A Comparison with the United States" (in Japanese), RIETI Discussion Papers, 02-J-018, Research Institute of Economy, Trade and Industry, November.

Nicoletti, G., S. Scarpetta and O. Boylaud (1999), "Summary Indicators of Product Market Regulation with an Extension to Employment Protection Legislation", OECD Economics Department Working Paper No. 226, Paris.

OECD (2001a), *The New Economy: Beyond the Hype,* Paris.

OECD (2001b), "Productivity and firm dynamics", *OECD Economic Outlook,* No. 69, Paris, June.

OECD (2001c), *Innovation and Productivity in Services,* Paris.

OECD (2002a), *Measuring the Information Economy 2002, www.oecd.org/sti/measuring-infoeconomy*

OECD (2002b), "Non-tariff Barriers in the ICT Sector: A Survey", TD/TC/WP(2001)44/FINAL, OECD, Paris, September.

OECD (2002c), *Purchasing Power Parities and Real Expenditures, 1999,* Paris.

OECD (2002d), "Productivity and Innovation: The Impact of Product and Labour Market Policies", *OECD Economic Outlook,* No. 71, June, pp. 171-183, Paris.

OECD (2003a), *OECD Communications Outlook 2003,* Paris.

OECD (2003b), *The Sources of Economic Growth in OECD Countries,* Paris.

Oliner, S.D. and D.E. Sichel (2002), "Information Technology and Productivity: Where Are We Now and Where Are We Going?", *Federal Reserve Bank of Atlanta Economic Review*, third quarter, pp. 15-44.

Oulton, N. (2001), "ICT and Productivity Growth in the United Kingdom", Working Paper No. 140, Bank of England, London.

Parham, D., P. Roberts and H. Sun (2001), "Information Technology and Australia's Productivity Surge", Staff Research Paper, Productivity Commission, AusInfo, Canberra.

Pilat, D., F. Lee and B. Van Ark (2002), "Production and use of ICT: A sectoral perspective on productivity growth in the OECD area", *OECD Economic Studies*, No. 35, Paris, forthcoming.

Sabourin, D. (2001), "Skill Shortages and Advanced Technology Adoption", Working Paper No. 175, Microeconomics Analysis Division, Statistics Canada, Ottawa.

Scarpetta, S., A. Bassanini, D. Pilat and P. Schreyer (2000), "Economic Growth in the OECD Area: Recent Trends at the Aggregate and Sectoral Levels", OECD Economics Department Working Paper No. 248, Paris.

Scarpetta, S., P. Hemmings, T. Tressel and J. Woo (2002), "The Role of Policy and Institutions for Productivity and Firm Dynamics: Evidence from Micro and Industry Data", OECD Economics Department Working Papers, No. 329, OECD, Paris.

Schreyer, P. (2001), "Computer Price Indices and International Growth and Productivity Comparisons", OECD Statistics Working Paper, OECD.

Schreyer, P., P.E. Bignon and J. Dupont (2003), "OECD Capital Services Estimates: Methodology and a First Set of Results", OECD Statistics Working Paper, Paris, forthcoming.

Simon, J. and S. Wardrop (2002), "Australian Use of Information Technology and Its Contribution to Growth", Research Discussion Paper RDP2002-02, Reserve Bank of Australia, Sydney, January.

Solow, R.M. (1987), "We'd Better Watch Out", *New York Times*, July 12, Book Review, No. 36.

Stolarick, K.M. (1999a), "IT Spending and Firm Productivity: Additional Evidence from the Manufacturing Sector", CES WP-99-10, Center for Economic Studies, Washington, DC.

Stolarick, K.M. (1999b), "Are Some Firms Better at IT? Differing Relationships between Productivity and IT Spending", CES WP-99-13, Center for Economic Studies, Washington, DC.

Tachibana, T. (2000), "The Survey on ICT Usage and E-Commerce on Business in Japan", paper presented 2000 at the Voorburg Group on Services Statistics meeting, Madrid, 18-22 September.

Triplett, J.E. (1999), "The Solow Productivity Paradox: What Do Computers Do to Productivity", *Canadian Journal of Economics*, Vol. 32, No. 2, pp. 309-334.

Triplett, J.E. and B.B. Bosworth (2002), "Baumol's Disease Has Been Cured: IT and Multifactor Productivity in US Services Industries", paper prepared for Brookings workshop on services industry productivity, Brookings Institution, Washington, DC, final version, 18 September.

United States Council of Economic Advisors (2001), *Economic Report of the President, 2001*, United States Government Printing Office, Washington, DC, February.

Van Ark, B., J. Melka, N. Mulder, M. Timmer and G. Ypma (2002), "ICT Investments and Growth Accounts for the European Union, 1980-2000", Research Memorandum GD-56, Groningen Growth and Development Centre, Groningen, *www.eco.rug.nl/ggdc/homeggdc.html*

Van Ark, B., R. Inklaar and R.H. McGuckin (2002), "Changing Gear Productivity, ICT and Services: Europe and the United States", Research Memorandum GD-60, Groningen Growth and Development Centre, Groningen, *www.eco.rug.nl/ggdc/homeggdc.html*.

Van Der Wiel, H. (2001), "Does ICT boost Dutch Productivity Growth", CPB Document No. 016, CPB Netherlands Bureau of Economic Policy Analysis, December.

Vickery, G. and J. Northcott (1995), "Diffusion of Microelectronics and Advanced Manufacturing Technology: A Review of National Surveys", *Economics of Innovation and New Technology*, Vol. 3, No 3-4, pp. 253-275.

OECD PUBLICATIONS, 2, rue André-Pascal, 75775 PARIS CEDEX 16
PRINTED IN FRANCE
(92 2003 03 1 P) ISBN 92-64-10128-4 – No. 53031 2003